Cerebral and Cerebellar Cortex – Interaction and Dynamics in Health and Disease

Edited by Stavros J. Baloyannis

Published in London, United Kingdom

IntechOpen

Supporting open minds since 2005

Cerebral and Cerebellar Cortex – Interaction and Dynamics in Health and Disease
http://dx.doi.org/10.5772/intechopen.87441
Edited by Stavros J. Baloyannis

Contributors
Abdullah Abdulrhman Al Abdulgader, Mario Díaz, Raquel Marin, Neli Cvetanova Vasileva, Jivko Dimitrov Jekov, Ichiro Nakajima, Ohba Hiroiku, Shinohara Mitsuyo, Mubarak Muhammad, Tasneem M. Hassan, Kenneth J. McLeod, Stavros J. Baloyannis

First published in London, United Kingdom, 2021 by IntechOpen
IntechOpen is the global imprint of INTECHOPEN LIMITED, registered in England and Wales, registration number: 11086078, 5 Princes Gate Court, London, SW7 2QJ, United Kingdom
Printed in Croatia

British Library Cataloguing-in-Publication Data
A catalogue record for this book is available from the British Library

Additional hard and PDF copies can be obtained from orders@intechopen.com

Cerebral and Cerebellar Cortex – Interaction and Dynamics in Health and Disease
Edited by Stavros J. Baloyannis
p. cm.
Print ISBN 978-1-83968-093-9
Online ISBN 978-1-83968-094-6
eBook (PDF) ISBN 978-1-83968-098-4

For EU product safety concerns:
IN TECH d.o.o., Prolaz Marije Krucifikse Kozulić 3, 51000 Rijeka, Croatia, info@intechopen.com or visit our website at intechopen.com.

Meet the editor

Stavros J. Baloyannis, Professor Emeritus of Neurology at Aristotelian University, Thessaloniki, Greece, graduated from the School of Medicine, Aristotelian University. He received training in neurology at Aristotelian University and Institute of Neurology, Queen Square, London. He also trained in neuropathology and electron microscopy at several institutions including the Institute of Neurology, Catholic University of Louvain, Belgium, University of Pennsylvania, USA, and Yale University, and in acoustic-neuropathology at Harvard University. He also trained in neuroimmunology at Yale University. His research interests include the blood-brain barrier, mitochondria in Alzheimer's disease, synaptogenesis, neurodegeneration, dendritic and synaptic pathology, and Golgi apparatus in dementias. Dr. Baloyannis' special interests include neuroethics, neurolinguistics, neurophilosophy, the history of neurosciences, neurology and art, and the brain and music. He is a member of 62 scientific societies and an honorary member of the Academy of Hellenic Air Forces. He is president of a society for the amelioration of quality of life in neurological diseases, and former president of an orthodox association for medical missions. He is a visiting professor at Tufts University, Massachusetts, USA, and Democritus University, Greece. He is the author of 29 textbooks, 727 papers in neurology and neurosciences, and 3 books of poems. From 1992 to 2011, Dr. Baloyannis was head of the Department of Neurology at the Aristotelian University, Thessaloniki, Greece where he is now director of the research institute for Alzheimer's disease.

Contents

Preface

The cerebral cortex is the most elegant and sophisticated structure of the brain. It plays a dominant role in elaborating sensory perception, organizing voluntary motion, memory, and judgment, emotions and behavior, interior life and thinking, speech and linguistic skills, art and music, learning and creativity, as well as the philosophy and the programming of life.

Personhood and social performance of human beings are developed, organized, and modulated by the cortex of the brain hemispheres in collaboration with the cerebellar cortex.

Each cortical area of the cerebral lobes plays a particularly crucial role in the creation, organization, and harmonization of mental faculties, in adjusting self-control and emotional stability, in developing self-identification and controlling multiple social interactions, shaping also the rational profile and the existential dimensions of a human being.

Amazingly, there is close cerebro-cerebellar connectivity and collaboration with the majority of the subcortical centers, concerning primarily the motor performances as well as most of the higher mental faculties, such as memory, emotions, perception, thinking transpersonal experiences, [1] creativity, innovation, and imagination.

The cerebellum plays a valuable coordinating role in the majority of the activities of the brain hemispheres [2]. Although the organization and uniform cytoarchitecture of the cerebellar cortex are very different from the considerable neuronal morphological variability of the cerebral cortex, the functional collaboration between cerebrum and cerebellum is a continuous harmonious process, resulting in the functional stability of the brain, [3] based on the contribution of the cerebellum in the consistency and appropriateness of motor and cognitive performances [4]. It is thought that the external information analyzed by the brain hemispheres is harmoniously matched with the internal predictions generated by the cerebellum [5].

A substantial body of evidence, based mostly on neuroimaging, advocates in favor of the important role that the cerebellum plays in the timing and adaptive manipulation of the majority of motor and cognitive processes generated and organized in the cortex of the brain hemispheres, including working memory, language processing [6], unconscious learning, music training, [7] and exploration of novel patterns of cognition and social behavior [8].

In pathological conditions, the cerebrum and cerebellum may continue their close functional collaboration via neuronal plasticity and cerebro-cerebellar reserve [9, 10]. However, in cases where the underlining etiology of the pathological alterations is severe, such as in toxic conditions, [11] viral diseases [12] affecting the cerebrum or the cerebellum, hereditary ataxias, cerebrovascular diseases, cortical degenerations, Alzheimer's disease, and other dementias, [13] the disorganization of the neuronal networks in the brain and the cerebellum induces marked alterations in mental performance, characterized by a serious cognitive decline as well as dysmetria of thought [14] and Schmahmann's syndrome [15] as a result of the wide disruption of the homeostatic equilibrium of the mind.

Cerebral and Cerebellar Cortex – Interaction and Dynamics in Health and Disease discusses several important issues of cerebro-cerebellar collaboration and interactions, such as the role of the cerebral and cerebellar cortex on consciousness and the orchestration of human consciousness from beyond the brain. In addition, the book analyzes dynamic praxis and spatial postural praxis in children 4–6 years old with typical development using neuropsychological methods to reveal that age is an important factor in the qualitative changes in the motor skills of children. The book also describes and discusses pathological alterations of the brain due to ischemic lesions and the mechanisms of brain damage. Reasonably, recovery from stroke based on neuronal plasticity and neosynaptogenesis would substantially ameliorate the quality of life of the patients. Thus, therapeutic interventions to enhance neuroplasticity would optimize the prognosis of the post-stroke condition of the patients, maximizing recovery from cerebrovascular episodes.

In the field of motor mechanisms and performances, a detailed study of the movement-related cortical potential (MRCP) for jaw movements in patients who underwent jawbone excision revealed that the brain's motor preparation process depends mainly upon a feed-forward system. The disruption of that system would affect seriously the information processing of the brain.

In cases of dementia due to Alzheimer's disease, the morphological alterations of the mitochondria in the cerebral and cerebellar cortex are among the initial phenomena in the broad spectrum of neuropathological changes. Therapeutic strategies protecting the mitochondria in the initial stages of Alzheimer's disease might be beneficial, leading to an escape from the tragic labyrinth of the disease.

In addition, cerebral perfusion plays a fundamental role in the performance of the cognitive function. A decrease of cardiac output associated with asymptomatic postural hypotension may be among the main etiological factors of cognitive aging, given that cardiac output is dependent on venous return. A technique applying soleus muscle stimulation, which would increase the soleus muscle pump function, may increase the venous return of the blood, improving cardiac output and anticipating cognitive aging.

In the enigmatic and unclear etiopathology of Alzheimer's disease, lipid rafts may also have a place as potential causative factors, participating in the amyloidogenic process of the amyloid precursor protein, given that in the initial stages of the disease a substantial lipid raft destabilization can be detected.

This book supports the concept of the close functional unity and harmonization of the brain and the cerebellum, underlining the important role that the cerebellar cortex plays in the performance of higher mental faculties, including creativity, [16] emotional processes, and homeostatic equilibrium of the human body [17].

We extend our gratitude to the authors for their excellent contributions.

Stavros J. Baloyannis
Professor Emeritus,
Departmet of Neurology,
Aristotelian University,
Thessaloniki, Greece

References

[1] Vandvert L. How the cerebellum and cerebral cortex collaborate to compose fractal patterns underlying transpersonal experience. A Fractal Epistemology for a Scientific Psychology: Bridging the Personal with the Transpersonal 2020; 38: 372.

[2] Van Essen DC, Donahue C J, Glasser MF. Development and evolution of cerebral and cerebellar cortex. Brain, behavior and evolution 2018;91: 158-169.

[3] Buckner RL, Krienen FM, Castellanos A, Diaz JC, Yeo BT. The organization of the human cerebellum estimated by intrinsic functional connectivity. J Neurophysiol. 2011;106:2322-2345.

[4] Schmahmann JD. An emerging concept. The cerebellar contribution to higher function. Arch Neurol. 1991;48(11):1178-1187.

[5] Van Overwalle F, Mariën P. Functional connectivity between the cerebrum and cerebellum in social cognition: a multi-study analysis. NeuroImage 2016;124: 248-255.

[6] Chein JM, Ravizza SM, Fiez JA. Using neuroimaging to evaluate models of working memory and their implications for language processing. Journal of Neurolinguistics. 2003; 16: 315-339.

[7] Vandervert L. How music training enhances working memory: a cerebrocerebellar blending mechanism that can lead equally to scientific discovery and therapeutic efficacy in neurological disorders. Cerebellum and ataxias 2015; 2(1): 1-10.

[8] Vandervert L,. Schimpf PH, Hesheng Liu. How working memory and the cerebellum collaborate to produce creativity and innovation. Creativity Research Journal 2007; 19(1): 1-18.

[9] Serra L, Gelfo F. What good is the reserve? A translational perspective for the managing of cognitive decline. Neural Regen Res. 2019;14:1219-1220.

[10] Mitoma H, Buffo A, Gelfo F, Guell X, Fucà E, Kakei S et al. Consensus paper. Cerebellar reserve: from cerebellar physiology to cerebellar disorders. The Cerebellum, 2020; 19(1):131-153.

[11] Abdallah M, Zahr N M, Saranathan M, Honnorat N, Farrugia N., Pfefferbaum A.et al. Altered Cerebro-Cerebellar Dynamic Functional Connectivity in Alcohol Use Disorder: a Resting-State fMRI Study. The Cerebellum.2021; 1-13. doi: 10.1007/s12311-021-01241-y

[12] Wang H, Li R, Zhou Y. et al. Altered cerebro-cerebellum resting-state functional connectivity in HIV-infected male patients. J. Neurovirol. 2018 ;24 : 587-596.

[13] Chen Y, Landin-Romero R, Kumfor F, Irish M, Hodges JR, Piguet O. Cerebellar structural connectivity and contributions to cognition in frontotemporal dementias. Cortex. 2020;129:57-67.

[14] Schmahmann JD. Dysmetria of thought. Clinical consequences of cerebellar dysfunction on cognition and affect. Trends Cogn. Sci. 1998; 2:362-370.

[15] Manto M, Mariën P.. Schmahmann's syndrome—identification of the third cornerstone of clinical ataxiology. Cerebellum Ataxias 2015; 2:2.

[16] Coolidge F L. The role of the cerebellum in creativity and expert stone knapping. Adaptive Behavior 2021;29 (2): 217-229.

[17] Cao H, Cannon TD. Cerebellar Dysfunction and Schizophrenia: From "Cognitive Dysmetria" to a Potential Therapeutic Target. Am J Psychiatry 2019; 176(7): 498-500.

Section 1

Physiology and Function

Chapter 1

Human Consciousness: The Role of Cerebral and Cerebellar Cortex, Vagal Afferents, and Beyond

Abdullah Abdulrhman Al Abdulgader

Abstract

Human Consciousness is one of most elusive issues in the scientific history. Its nature created major historical debate started thousands of years ago and still ongoing. Despite the explosive developments in the last 6 decades to explore its nature, the knowledge about it is still deficient. The important advances in the twentieth and 21st centuries in understanding cerebral cortex dynamics fortified by the dominant materialistic philosophical approach of the era dictated its impact on consciousness science, which is understood as sole human brain function. This chapter is a call for holistic perception of human consciousness incorporating the ancient wisdom of the human civilizations with the massive current advances in different disciplines of applied sciences. The description of René Descartes in the 17th century of the Cartesian dualism is timely to revisit with new holistic perspective, in view of the major advances of our understanding of heart brain communications, astrophysical resonances with, human heart and central nervous system frequencies, and signaling between humans and their large environment. Neural and psychological correlates of human consciousness which dominate the consciousness research nowadays should undergo revolutionary conceptual understanding to perceive consciousness as a massive universal event expanding from human genes to galaxies with cerebral cortex as major player.

Keywords: human consciousness, solar geomagnetic activity, heart based resonant fields theory of consciousness

1. Introduction and definition

Consciousness remains the hall mark defining human intelligence and interactive life and the true demarcation line between being and not being. In spite of being the most practical experience of self identity and intelligent life reactions in this life that we live, its nature remains an area of great debate and sometimes conflicting opinions between philosophers, biologist and intellectualists since the dawn of human scientific history. The twentieth century is known as the century of brain as there was exaggerated materialistic inflation of brain role in human functions but mainly consciousness. There is compelling scientific and rational evidence to convince scientific communities that the nature of consciousness involves dynamics inside the skull but essentially much beyond it in extreme dimensions between the skull and the sky. In addition to discussing the sophisticated neurobiological dynamics within the cerebral cortex, the main aim of this chapter is to open channels for holistic perception and understanding of human consciousness incorporating other scientific disciplines like the central role of human heart contribution to consciousness, quantum physics, as well as astrobiologigcal aspects of consciousness are going to be discussed.

2. Consciousness definition

2.1 The challenges

Since the dawn of humanity the ability of human beings to be alert, responsive and behave intelligently with emotions and identity were the subject of huge concerns in the philosophical, medical, psychological and religious communities. The explosive nature of diagnostic modalities in neuroimaging, medical physics and neurocardiology since world war 2 but more specifically in the last 20 years created revolutionary perspective of our understanding of the nature and origin of consciousness. Those advances were paralleled with numerous publications and selective conferences concerned with the brain and mind. We established unique conceptual congress, the King of Organs for Advanced Cardiac Sciences where heart and brain communications were discussed in unconventional ways in five international conferences (2006,2008,2010, 2012 and 2019) founded and chaired by the author of this chapter. One of the most challenging controversial and still ongoing scientific issues is the debate on how to define consciousness. The words *conscious* and *consciousness* is antique but appeared first in the documented English literature in the17th century followed by the world *self-conscious* and *self-consciousness* [1]. Consciousness is frequently used in different cultures and writings interchangeably with the Mind. In our understanding the term "Mind" is best preserved for psychological states and processes that might or might not be 'conscious'. In addition it is crucial to emphasize that mind and brain are not synonyms. Brain is structural correlates of the mind while mind is functional correlates of the brain. In similar way the tem knowledge is not synonym for consciousness as knowledge can be unconscious, or implicit.

Comprehensive understandings of scientific etymology demands the linguistic power in combination with the practical meanings as it is accepted and understood in the mainstream language. In this regard we suggest approaching consciousness with broad visionary perspective. For this reason we will define consciousness by referring to three major domains: First: the state of alertness and being vigilant, the opposite of which is coma as measured by Glasgow Coma Scale. This is predominantly of neurological nature. Second is the experience or the content of experience from time to time or 'what something looks like' and the inward connotation and feelings. This is predominantly of philosophical nature. Third is referred to the mental state with propositional content like fear, anger or appreciation. Most research in medical literature has natural tendency to neglect this third meaning of consciousness. This state of continuous historical uncertainty and debate about consciousness is in our opinion justified because of lack of knowledge of the origin, dimension and fate of our current life consciousness. The fact that the nature of consciousness cannot be explained as deduction from pathological alterations in the brain led to the fact that the mysterious mission of understanding human consciousness will be impossible without involving dimensions out of cerebral cortex. It can be looked as property of highly complex dependent biological systems which is adaptive, and highly interconnected.

The phenomena of access consciousness where information are accessed to the brain from different energetic cosmic levels is a major gate to explore in the comprehensive science of consciousness although by itself, it is unconscious. What increase complexity is the historical believe in different civilizations and religions that consciousness will never disappear but transform from one realm to another. Recently consciousness research refers to the **Consciousness Tetrad** which describes four escalating levels of consciousness starting from the **default consciousness** which is the default state that separates the living from the dead, rising to the **aware consciousness** which looks at consciousness as a continuum of states ranging from awake to sleep to drowsiness to semiconscious states like stupor and finally coma. The third level is the **operational consciousness** which is consciousness related to motor, sensory, cognitive, ethical, creative, emotive, and other abilities and capabilities and awareness of all mental operations. The fourth level is what philosophers refers to as the transcendence of human soul called the **exalted consciousness** where the most elite and spiritual experience of a human being can be achieved as the person is getting closer to the thresholds of the highest intelligence ever, who masters the emergence and fate of death and life and all creatures and universes. Our mentor and ex colleague and fellow member in the scientific advisory board of HeartMath Institute the famous neuroscientist the late Karl Pribram simplify the current deficient perception of consciousness saying that there is little c consciousness refed to being awake or sleep and big C consciousness referring to the intelligence and information that organizes the universe and all that is within it.

2.2 Definition of consciousness

Agreement in definition and overlapping terms is important to navigate safely and target the phenomenology of human consciousness as precisely as possible. To recapitulate the wide spectrum of meanings and domains the author define Consciousness as *a state of alertness and being aware, active and vigilant of the self and surroundings with volition, based in memory and personal identity.* **This state is** *ineffable* **and** *intrinsic* **and express itself in presence of soul through activation of different sensing and perceptive body organs but may pass through lighter densities and variable dimensions in quantum nature, if soul leaves the body**.

The following discussions in this chapter will expand the understanding in those directions.

3. Neurobiological basis of consciousness

The level of human consciousness is the collective activity of widespread areas of bilateral association of cortical and subcortical structures and possibly other interconnected biological and astrophysical systems. Due to the complex nature of consciousness origin and dimensions, it would be too artificial and nonrealistic to confined consciousness discussion in cortical, subcortical dimensions as is the product of interaction and connections of complex biological and non-biological networks.

Although much has been learned about the neuroanatomical structures participating in consciousness, there is always great demand to establish the physiological and physical mechanisms through which consciousness is generated in these networks, structures and dimensions. In this section we are going to elaborate with some details about "brain related consciousness systems" including cortical components, as well as subcortical components comprising multiple parallel arousal systems in the upper brainstem, thalamus, hypothalamus, and basal forebrain, cerebellum as well as sensory and motor components and neurotransmitter pathways, interaction of which will determine the amount of alertness, attention, and awareness.

Cortical components consisting medially of the medial frontal, anterior cingulate, posterior cingulate, and medial parietal (precuneus, retrosplenial) cortex. On the lateral surface, it includes the lateral frontal, anterior insula, orbital frontal, and lateral temporal–parietal association cortex. The major subcortical networks that regulate level of consciousness including the thalamus and subcortical arousal nuclei acting through multiple neurotransmitters (glutamate, acetylcholine, gamma amino butyric acid (GABA), norepinephrine, serotonin, dopamine, histamine, orexin) that arise from the upper brainstem, basal forebrain, and hypothalamus are going to be discussed.

The consciousness experience remains more complex than simple understanding of possible structure or network functions. The content of consciousness at certain time period is interdependent on the substrate of structure(s) and network(s) activated during that time to yield the specific conscious experience as will be discussed in this section.

3.1 The correlates of consciousness

In spite of the developments in the field of consciousness in the last two decades it is not clear how any physical process, such as neural activity, can give rise to a subjective phenomenon such as conscious awareness of an experience. For this reason, very important observation for researchers in the field of neurobiology of consciousness is to know that the causal relationship of the objective detection of neuronal activation and the subjective awareness of conscious experience is uncertain. Neuroscientist suggested the idea of the Neuronal Correlates of Consciousness (NCC) to be able to study the possible minimal model or the smallest possible building components of conscious percept or explicit memory.

Gamma-band oscillations around 40 Hz is proposed as the band that correlate conscious processing. Examples of NCC documented in literature are: the intralaminar nuclei of the thalamus [2], re-entrant loops in the thalamocortical systems [3], an extended reticular thalamic activation system [4], neural assemblies bound by N-Methyl-D-Aspartate (NMDA) [5], the inferior temporal cortex [6], visual cortex connections to the prefrontal cortex [7, 8] and visual processing within the visual stream [9]. With the advances of both behavioral sciences and computational engineering consciousness scientists, innovate are under investigations to discover ways to couple Neuronal Correlates of Consciousness (NCC) with Behavioral Correlates of Consciousness (BCC) and Computational Correlates of Consciousness (CCC) in order to promote our understanding of more comprehensive perspective of human consciousness. Building conscious machine through the new understanding of computational models [10] artificial intelligence and cognitive robotics is pursed.

3.2 Ascending reticular activating system and consciousness (ARAS)

Revising history of arousal in modern medicine document (ARAS) as one of the first described structures responsible of enhanced arousal [11]. After decades of researchers efforts we know that what was described as (ARAS) is not a structure of brain stem nuclei per se but is a group of specialized nodes in a complex network and pathways that controls arousal. This network includes the cholinergic nuclei in the upper brainstem and basal forebrain, The posterior hypothalamus histamine projection, and noradrenergic nuclei, especially the locus coeruleus. The dopamine and serotonin pathways that arise from brain stem are thought to be part of (ARAS). The thalamus which constitute crucial synaptic relay for most sensory and intracerebral pathways is located strategically at the apex of (ARAS) and have mediated major control on most of its activities [12, 13]. Thalamic burst discharges are generated through extensive inhibitory axon collaterals, produced by special thalamic, ARAS coordination. Those discharges are responsible for gating specific reticular information which is in turn transmitted back to the cortex, and this reverts the information back to the brainstem [14]. Positron emission tomography (PET) investigation during slow-wave sleep [15] and anesthesia [16] documented selective thalamic and ARAS hypometabolism through studying functional neuroimaging of normal human sleep and studying the neurophysiologic basis of anesthetic induced unconsciousness.

3.3 Amygdala contributions to consciousness

Amygdala, the brain's center for emotions, occupies major position in the neurology and biology research concerned with working memory, long-term memory, and attention. It is strongly linked with social interactions region in the brain, namely, the orbital cortex. Tight tripartite network constitutes robust pathways from amygdala connected to neurons in the thalamus which in turn connect directly to the orbital cortex. The pathways from the amygdala to the orbital cortex and to the thalamus are dual and distinct by function, morphology, neurochemistry [17]. This highly sophisticated and specialized pathways provide strong evidence that emotions influence higher cortical areas concerned with affective reasoning. In addition, Investigating the neurobiological bases of executive functions suggest that amygdala facilitates cognitive performance during challenging tasks between the amygdala and cognitive systems. For this reason neurotransmitters like dopamine and noradrenaline may contribute important role between the amygdala and higher cognition [18, 19]. In our opinion, the well-established role of amygdala in emotions and the additional relation to cognition are both integral to each other and support the establishment of comprehensive intelligent emotional model as a cornerstone of human consciousness experience.

3.4 The cerebellum and its contributions to consciousness

Functions related to movement, gait, posture and balance were the traditional functions related to cerebellum. In the last two decades cerebellum was found to have regulatory functions concerned with emotion processing, cognition, behavior, and collectively consciousness experience [20, 21]. The cognitive role of the cerebellum can be understood by looking at its afferent and efferent connections. The most important of the central afferent circuits is the corticopontocerebellar pathway which emanates from the motor and sensory cortical areas. The pontocerebellar tracts connect with the pontine nuclei then it connect with the contralateral cerebellar hemisphere in a somatotopic manner -which denotes feeling or consciousness experience- of point to point correspondence of an area of the body to a specific point on the cortex. Peripheral cerebellar pathways originate from the brainstem. Via the red nucleus and ventrolateral nucleus of the thalamus, the cerebellum exerts most of its output to the brain stem and the cerebral motor cortex [22]. Efferent cerebellar pathways are four and ultimately connects to the following critical structures: pons, medulla oblongata a, reticular formation, basal ganglia, corticospinal and reticulospinal pathways and limbic cortices (cingulate and parahippocampal gyri). Those sophisticated networks and connections of afferent (corticopontocerebellar) and efferent (cerebellothalamocortical) pathways, the cerebellum can exert highly complex regulatory role and integrate information to the cortical cerebral areas related to cognition and ultimately the consciousness experience [23]. The ongoing collective data from different discipline in genetics, neuropsychological research, structural and functional brain imaging studies will provide better perspective of the integral role of cerebellum in consciousness [22].

3.5 The thalamus and its contribution to consciousness

Thalami are pair of large ovoid organs that form most of the lateral walls of the third ventricle of the brain in humans. Thalamic main nuclear divisions and nuclei are: midline thalamic nuclei, anterior nuclear group, medial nuclear group (mediodorsal nucleus), lateral nuclear group, thalamic reticular nucleus and intralaminar nuclei. Nearly all information directed to the cortex first reaches the thalamus. The thalamus transmits this information and reciprocally receives an even greater number of connections back from the cerebral cortex. For this reason, the thalamus is considered as a major player in all forebrain functions including consciousness. The thalamus relays the content of consciousness, and also controls its level via specialized circuits that act as regulator of arousal level and are critical for selective attention. The specific thalamic relay nuclei communicate with the cerebral cortex regarding each sensory and motor function. For this reason the thalamus with its extensive nuclei connections is thought to be responsible for all the individual contents of consciousness [24]. Corticothalamic rhythms are thought to be generated by The reciprocal connections between thalamic relay nuclei and the thalamic reticular nucleus during normal sleep and waking activity, as well as in pathological rhythms such as epilepsy [25]. the intralaminar thalamus plays an important role in transmitting arousal influences from strategic location, namely, the midbrain and upper pontine cholinergic and glutamatergic systems to the cortex. *Lesions in this crucial area of the upper pons and midbrain produce deep coma, whereas in comparison lesions in the lower pons or medulla oblongata do not typically disrupt consciousness.*

3.6 The frontoparietal circuits and its role in consciousness

The contribution of frontoparietal activity to conscious perception was suggested by neuroimaging studies. In addition to visual perception due to activity in the ventral visual cortex, the parietal and prefrontal areas contribution seems to be essential for awareness [26, 27]. The network nodes for correlates of consciousness are thought to be divided to primary and secondary. Early activity in the occipital lobe correlates with the perceptual processes, which is detrimental for later process, namely, the activity in the frontoparietal areas. Access consciousness, in comparison to the phenomenal subjective consciousness due to mainly activation of sensory regions, refers to the direct control of experience through reasoning, reporting, or action. This type of higher functioning needs the involvement of the frontoparietal areas [28].

3.7 The prefrontal cortex (PFC) and consciousness

The PFC constitutes a large portion of the frontal lobe that includes most of the cortical tissue anterior to the central sulcus which can be divided to five main areas. The role of the prefrontal cortex (PFC) as an NCC is the source of debate between prefrontal theories and posterior theories of consciousness. The strongest argument point of posterior theories of advocates is the preservation of consciousness in patients with PFC lesions. Apparently, they limit their definition of consciousness to the state of alertness and vigilance, which is too deficient definition. In our view, adopting our comprehensive definition of consciousness, mentioned earlier, will make both conflicting parties complementary rather than competitive. The five main areas of the PFC –namely The anterior prefrontal cortex the caudal prefrontal cortex; the dorsolateral prefrontal cortex; the ventrolateral prefrontal cortex; and the medial prefrontal cortex -are extensively connected with sensory areas, which conceivably denotes that PFC is essential part of the consciousness experience although the exact mechanism of how the sensory information could become conscious is still not well understood. NCCs involving PFC might be subtle neurological activity. The fact that common neuroimaging techniques are not sufficiently sensitive to detect subtle differences in neural activity should be considered in future research discussing the role of OFC in consciousness [29].

3.8 Precunues, posteromedial parietal lobe and consciousness

Precumues or the mesial extent of Brodmann's area is a cortical region located in the posteromedial portion of the parietal lobe. It is well known with its widespread connections with both cortical and the subcortical structures. Recent Functional imaging findings in healthy subjects suggest a central role for the precuneus in a wide spectrum of higher functions, including visuo-spatial imagery, episodic memory retrieval and self-processing operations. Precuneus and surrounding posteromedial areas are among the most hot spots of the brain as it is displaying high resting metabolic rates. It is characterized by transient decreases in the tonic activity during engagement in non-self-referential goal-directed actions [28]. It is thought that precuneus is involved in the interwoven network of the neural correlates of self-consciousness, engaged in self-related mental representations during rest. The evidence is supportive of the involvement of precunues in the endogenous signaling function during conscious resting state. This hypothesis is consistent with the selective hypometabolism in the posteromedial cortex reported in a wide range of altered conscious states, such as sleep, drug-induced anesthesia and vegetative states [30].

3.9 Consciousness related neurotransmitter systems and pathways

3.9.1 Glutamatergic arousal systems

The most prevalent excitatory neurotransmitter in the central nervous system is Glutamate. It functions seems to be critical in initiation and maintaining of sleep and wakefulness. Arousal system pathways arising from the midbrain and upper pontine reticular formation that project to the thalamus and basal forebrain as well as the widespread projections from the thalamic intralaminar nuclei to the cortex are thought to be mediated by glutamate [31]. Through interaction with other types of neurons, the glutamatergic neurons can regulate sleep stages. With this type of arrangement, complex sleep–wake regulation network in the brain is made [32].

3.9.2 Cholinergic arousal systems

Acetylcholine, although being, the major neurotransmitter of the peripheral nervous system, plays neuromodulatory function in the central nervous system (CNS). The brainstem pontomesencephalic reticular formation and the basal forebrain are the two main sources of cholinergic projections neurons in CNS. Brain stem arousal is thought to act in a synergistic manner with the noncholinergic putative glutamatergic pontomesencephalic neurons which project to intralaminar thalamus and basal forebrain [24, 31, 33]. The brainstem and basal forebrain cholinergic systems work together to abolish The cortical slow wave activity which is known to be enhanced with pathological brain function as in stroke, schizophrenia, depression, Morbus Alzheimer, and post-traumatic stress disorder are abolished by the brainstem and basal forebrain cholinergic systems and this ultimately will promote an alert state [31, 34] Muscarinic acetylcholine receptors are the major receptor type operating in cholinergic arousal in the CNS, although nicotinic receptors may also play an important role [24]. The result of pharmacological blockage of cholinergic neurons in the CNS can be deduced from its functional areas connections, resulting in acute state of delirium and memory loss. In the contrary, the miracle of human brain creation is shown in the preservation of consciousness with experimental selective damage to cholinergic neurotransmission [35]. This can be explained by the multiple parallel neurotransmitter systems are participating in maintaining the consciousness.

3.9.3 GABAergic arousal systems

The most prevalent inhibitory neurotransmitter in the CNS is GABA. It is known for its major role in regulating arousal. Several long-range GABAergic projection systems also contribute to controlling arousal. Arousal is promoted by some GABAergic neurons in the basal forebrain as these inhibitory neurons in turn project to cortical inhibitory interneurons [24, 36]. On the other hand the overall effects of basal forebrain GABAergic neurons on arousal process is variable with variable firing patterns on cortex and sleep awake cycle. Long GABAergic projections have their overall function as inhibitors for arousal process. These include neurons like ventral lateral preoptic nucleus which is known for its widespread inhibitory projections to almost all subcortical arousal systems [37]; forebrain and hypothalamus inhibitory neurons namely lateral septal GABAergic neurons [38]; and the GABAergic containing neurons nucleus namely the thalamic reticular nucleus that is projecting to the remainder of the thalamus and projecting to the brainstem reticular formation [39]. Regions of the thalamus including the intralaminar nuclei are inhibited by GABAergic neurons in the globus pallidus internal segment. It is thought that the inhibition of the globus pallidus to remove the tonic inhibition of the intralaminar thalamus with medications like zolpidem in minimally conscious state, or benzodiazepines in catatonia is the possible mechanism for the paradoxical arousal effects of those GABA agonist [24, 40]. The loss of consciousness in partial seizures is thought to be due to activation of these multiple GABAergic inhibitory projections converging on the subcortical arousal.

3.9.4 Noradrenergic arousal systems

In proximity to the fourth ventricle, in the rostral pons the locus ceruleus contains the norepinephrine (noradrenaline) neurons. Inhibition of locus ceruleus neurons with drugs like selective α-2 agonists such as clonidine or the anesthetic agent dexmedetomidine is the possible mechanism of action yielding profound depression of arousal. In contrary selective blockage or removal of noradrenergic neurons will impair arousal but will not end up in deep coma. This can be explained, like the situation mentioned in cholinergic arousal systems, by multiple parallel neurotransmitter systems are participating in maintaining the consciousness. Norepinephrine neurons type are also found in the lateral tegmental area extending into the more caudal pons and medulla [24, 41]. Sleep–wake cycles, attention, and mood are regulated via ascending noradrenergic projections that reach the cortex, thalamus and hypothalamus. Modulation of autonomic nervous system function and pain gating is operated through descending projections to the brainstem, cerebellum, and spinal cord.

3.9.5 Serotoninergic arousal systems

The midline raphe nuclei of the midbrain, pons, and medulla contains most of the serotonergic neurons. Projections to the entire forebrain are received from the more frontal serotonergic neurons in the midbrain and upper pontine raphe nuclei, participating in regulation of sleep–wake cycle. Serotonergic systems occupies major position in psychiatric practice as dysfunction of which is thought to play a role in a number of psychiatric disorders including, anxiety, depression, obsessive–compulsive disorder, aggressive behavior, and eating disorders. Modulation of breathing, pain, cardiovascular system, temperature control,, and motor function is attributed to the caudal serotonergic neurons in the pons and medulla. The dorsal raphe and median raphe are thought to be the most important rostral raphe nuclei participating in arousal process [42]. The contribution of serotonergic neurons to the arousal process with either promotion or inhibition, is complex due to the wide diversity of serotonin receptors in different regions of the brain [43, 44]. The lifesaving arousal response to hypoventilation and high carbon dioxide tension is thought to be promoted by brainstem serotonergic neurons located rostrally [45].

3.9.6 Dopaminergic arousal systems

The substantia nigra pars compacta and the adjacent ventral tegmental area of the mid brain are the regions where dopaminergic neurons are mostly found. Three ascending dopaminergic projection systems will emanate from those nuclei projecting to vital cortical and subcortical regions with substantial contribution to consciousness process: (1) the mesostriatal (nigrostriatal) pathway (2) the mesolimbic pathway (3) the mesocortical pathway. Those three projections arise from substantia nigra (pathway 1) and ventral tegmental area (pathway 2 and 3) reaching to the caudate and putamen (pathway 1), limbic structures including the medial temporal lobe, amygdala, cingulate gyrus, septal nuclei, and nucleus accumbens (pathway 2), the prefrontal cortex and the thalamus (pathway 3). Dopamine can have dual effect on the thalamus and cortex either activation or inhibitory [46]. Schizophrenia related apathy and the reduction of motivation and initiative, seen in frontal lob pathologies, abulia, and akinetic mutism are thought to be due to impaired dopaminergic transmission to the prefrontal cortex [47].

3.9.7 Histaminergic arousal systems

In the posterior hypothalamusan an important nucleus is called tuberomamillary nucleus where most of the Histamine-containing neurons are found. In addition a few scattered histaminergic neurons can be seen in the midbrain reticular formation. The entire forebrain including cortex and thalamus receives extensive ascending projections emanating from the tuberomamillary nucleus, while the brainstem and spinal cord receives descending projections [48]. Anti-histamine medications are intended to act on peripheral histamine release from mast cells, but are well-known to induce drowsiness presumably through central actions (White and Rumbold, 1988). Anti histamine medications are thought to act centrally inhibiting the arousal function of histamine on cortex [49] and thalamus [50] resulting in drowsness. In addition other hypothalamic nuclei, the basal forebrain, brainstem cholinergic and noradrenergic nuclei may contribute to the arousal actions of histamine. Histamine effect is thought to be receptor specific as activation of H1 receptors will facilitate alertness where H3 receptors activation will result in drowsiness.

3.9.8 Orexinergic arousal systems

Orexin from *orexis,* means "appetite" in Greek; is a neuropeptide produced in neurons of the perifornical, lateral, and posterior hypothalamus that regulates appetitel, wakefulness, and arousal. It projects to cortex and almost all arousal subcortical systems. Alternatively, in some publications it is called hypocretin. Two research groups in rat brain discovered it in 1998 almost in the same time: Masashi Yanagisawa's lab at the University of Texas and Sakurai T et al. [51]. Deficiency of the orexin systems will result in, a disorder characterized by excessive daytime sleepiness and pathological transitions into rapid eye movement sleep namely, narcolepsy [24]. The beneficial effects of modafinil in preventing the symptoms of narcolepsy, shift work sleep disorder, and excessive daytime sleepiness associated with obstructive sleep apnea is thought to be through activation of orexin neurons.

3.9.9 Adenosine and arousal

Hydrolysis of Adenosine Mono phosphate (AMP) and S adenosyl- homocysteine (SAH) will result in adenosine production which is known as a somnogenic substance that has control on normal sleep–wake patterns. The neuroanatomical sources of adenosine are not well known, but functionally it is well known neuromodulator contributing to the conscious arousal The adenosine system can affect the gating of Slow Wave System-Slow Wave Activity expression. Adenosine affect is through modulating of the arousal level, thereby altering the duration of time during which sleep homeostasis and function can occur [52]. Adenosine receptor stimulation is expected theoretically to act as a potential treatment for insomnia. In spite of the fact that $A_{2A}R$ agonists strongly induce sleep, classical $A_{2A}R$ agonists have adverse cardiovascular effects that restrict its use clinically. In addition the passage of adenosine across the blood–brain barrier (BBB) is known to be poor with evidence of rapid degradation inside endothelial BBB cells. Infusing of selective $A_{2A}R$ agonist CGS21680 increases the release of GABA in the tuberomammillary nucleus (TMN), but not in the frontal cortex and decreases histamine release in the frontal cortex and medial preoptic area. Adenosine arousal effect can be blocked by coffee and theophylline.

4. Brain molecular and cellular events and the neuron firing, is it all the sole source of human consciousness

4.1 Refute of the 20th century doctrine on consciousness origin

The last 7 decades conceptual model of the consciousness scientific dilemma in general human knowledge as well as in scientific specialties in psychiatry, neurology, clinical neuroscience and all related disciplines was based on reductionist concepts that aimed at naturalizing all phenomena of mind including memory and other higher functions, to solely, cellular and molecular mechanisms of the human nervous system [53]. This dogma occupied the scientific understanding of the twentieth century. As a matter of fact those reductionist ideas as well as their opponents extended few thousands of years deep in the human history. Example of the opponents are the phrenologicals as documented by work of the Austrian anatomist Franz Joseph Gall (1758–1828), [54]. In fact in ancient Egyptian wisdom the role of human brain as the source of wisdom and consciousness was not of value. In fact, when creating a mummy, the Egyptians scooped out the brain through the nostrils and threw it away [55]. The ancient Egyptians believed that the heart, rather than the brain, was the source of human wisdom, as well as emotions, memory, the soul and the personality itself. The father of the reductionist theory of brain functioning in todays medicine is Wilder Penfield's (1891–1976) who adopt the concept that electrical stimulations in certain brain areas produce experiential phenomena [56]. The originality and innovative level of Penfield's contributions to the field of neurophysiological localization of the higher psychological functions in the human cortex as well as the purity of his operational research approaches was questioned and criticized. Now a days, Penfield approach with his neurological and psychiatric patients is of considerable academic debate in the scientific communities [57]. In historical appraisal R. Nitsch and F. W. Stahnisch in the journal Cerebral Cortex challenged Penfield original concept of experiential phenomena elicited by electrical Stimulation of the human cortex. They revisited Penfield clinical work and found that the actual results obtained from electrical stimulation studies of the brain are far less conclusive, than his firm assertions made during Penfield Gordon Wilson Lecture in 1950. They stated clearly "In-depth comparison with the original stimulation map shows clearly that the original stimulation protocol did not support this repetitive account by stimulation at the same point". There was no consistent response of defined experiential phenomena observed upon stimulation of an individual stimulation point of the original work. In addition there was no full memory repertoire could be elicited. Patient's stimulation records did not yield stream of an individual's consciousness [58]. The heaviness of the scientific evidence emphasizing that consciousness is a complex reconstructive process, not merely limited to electrophysiological stimulation and recordings is beyond the stage of simply overlooking the situation. ***There is compulsive stream of scientific power to depart from our current very limited perspective of the process of human consciousness to be limited to the box of the skull and to be expanded as far as the sky. It is justifiable to claim that neuronal reductionism is a failed theory and that the search for an answer to the question about the origin of consciousness has to take a novel turn.***

4.2 Consciousness without a cerebral cortex

The thalamocortical complex does not seem to be critically essential for consciousness experience. Brainstem mechanisms by its own can create adequate consciousness state. This means that Consciousness without a cerebral cortex is possible [59]. Penfield and Jasper note that a cortical removal even as radical as hemispherectomy deprived their patients certainly from of information and discriminative capacities but not consciousness [60]. An explicit reference to the midbrain reticular formation was always included in Penfield and Jasper definition of their proposed centrencephalic system. Sprague in 1966 contribute significantly to consciousness research after performing complete removal of the posterior visual areas of one hemisphere in the cat. Agrees well with the Penfield and Jasper perspective that without cognizance of potential subcortical contributions to cortical damage deficit, the cortical functions will be counterfactually inflated [61]. Striking scientific agreement arguing strongly against the necessity of cerebral cortex for consciousness experience is seen in children born without cortex, namely Hydranenecephalic children. It is a congenital anomaly of the brain where for genetic or acquired reasons the cerebral cortex is drastically under developed and replaced by cerebrospinal fluid (**Figure 1**).

Neurological evaluation reveals they are responsiveness to their surroundings and conscious. Personal observations reported by hundreds of families of affected children stressed on the fact that their responsiveness is most readily to sounds, but also to salient visual stimuli. To the surprise a paradox phenomena in this regard is rarity for any auditory cortex to be spared in those children in spite of their impressive sound responsiveness. Bjorn Merker wrote a unique chapter entitled "Consciousness without a cerebral cortex: A challenge for neuroscience and medicine" which appeared in Behavioral and Brain Sciences and was able to spent seven days of observation with 5 families in a visits to Disney World. He stated that "They express pleasure by smiling and laughter, and aversion by "fussing," arching of the back and crying (in many gradations), their faces being animated by these emotional states. The children respond differentially to the voice and initiatives of familiars, and show preferences for certain situations and stimuli over others, such as a specific familiar toy, tune, or video program, and apparently can even come to expect their regular presence in the course of recurrent daily routines. "[62] It is woeful that many medical institutes label hydranenecephalic children to be in a vegetative state. On the other hand *the absence of the cerebral cortical tissue with preserved consciousness as well as all normal mental functions and normal neuropsychological testing is not a myth*. In other words, normal human being can be seen conscious and mentally good without brain tissue. Lionel Feuillet et al., published in the Lancet 44 years old French man married, a father of two children, and worked as a civil servant with otherwise normal neurological development and medical history. History revealed ventricloatrial shunt surgery in childhood with two revisions after mild symptoms. He achieved intelligence quotient (IQ) of 75, verbal IQ

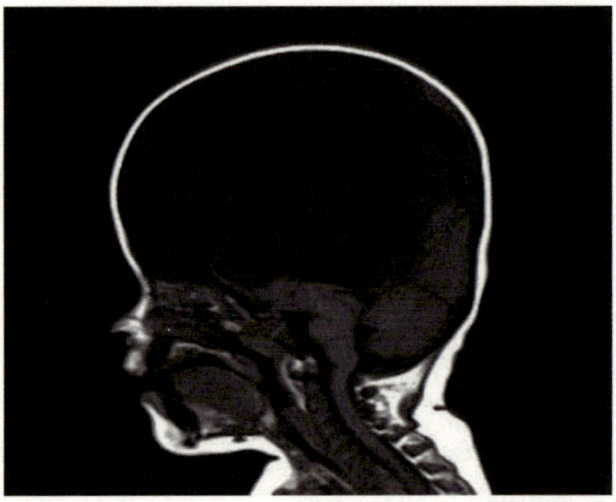

Figure 1.
Sagittal MRI section for a child demonstrating drastic underdevelopment of cerebral cortex with only remnants of occipital and temporal lobes. Cerebellum and brainstem are intact.

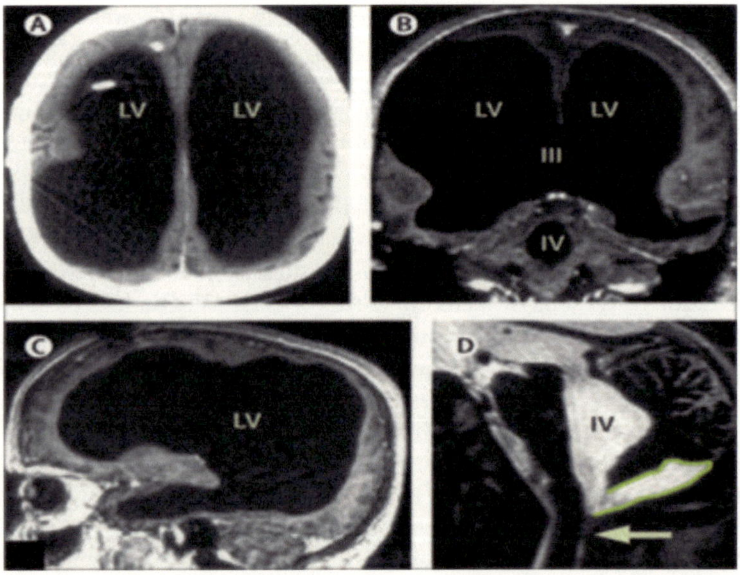

Figure 2.
44 years old french man with 90% absence of his cerebral cortex. His consciousness, mentality as well as social life were otherwise normal. LV=lateral ventricle. III=third ventricle. IV=fourth ventricle. Arrow=Magendie's foramen.

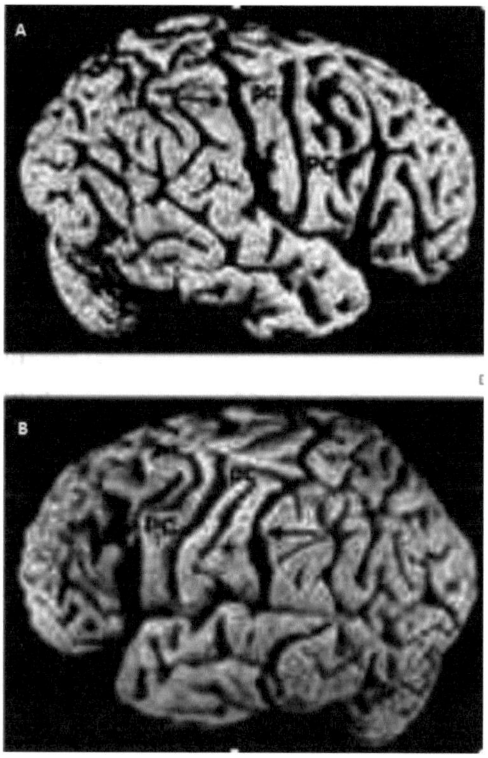

Figure 3.
Einstein's brain is no exception to the most common of patterns, showing (A): Typical posteriorly rising Sylvian fissure on the right (arrow) and (B): A parietal operculum on the left (asterisk).

was 84, and his performance IQ was 70. MRI revealed massive hydrocephalus with all brain ventricles, lateral, third, and fourth were enlarged. 90% of cerebral cortex was absent with remnants of very thin cortical mantle and a posterior fossa cyst (**Figure 2**) [63].

4.3 Structural brain components and human intelligence are they parallel?; the Einstein's brain

Thinking of intelligence from the point of computer and artificial intelligence language will denote the fact that the higher the capacity and intellectual power of a computer requires larger hard ware and more sophisticated computers. The comparisons is not valid in case of human brain as the anatomical study of the most intelligent human in the 20 century Albert Einstein's brain was not showing any convincing anatomical difference than any body brain. Witelson and colleagues' claim that Einstein's brain lacks a parietal operculum on the left and right sides. A M Galaburda from Harvard Medical School -and others- argues strongly against this and documented that Einstein's brain is no exception to the most common of patterns, showing a parietal operculum on the left and the typical posteriorly rising Sylvian fissure on the right (**Figure 3**) [64, 65].

5. Orchestration of the human consciousness from beyond the brain

It is conspicuous for the ingenious observer in the consciousness scientific arena that the inability to explain and match facts and observations and the failure to reproduce the exact consciousness experience incorporating current knowledge in the field implies presence of deficient rings in the long chain that demand more comprehensive perspective. In this regard we established the King of Organs International Congress for Advanced Cardiac Sciences and held five international congresses (2006,2007,2008,20,101,2012 and 2019). The King of Organs congresses are an international collaborative efforts between international renewed scientists in cardiac sciences, psychologists, astrophysicist, mathematicians, geologists, space engineers, signal analysis specialists and other related disciplines. It is chaired by us in Prince Sultan Cardiac Center (Alhasa, Saudi Arabia). Our academic partners are the HeartMath Institute and the Global Coherence Initiative (Boulder Creek, CA, USA), American Institute of Stress (NY, USA), The Global Consciousness Project (Institute of Noetic Sciences, USA), and other western and eastern reputable universities and collaborators. *Our mission is to decode the great mystery of consciousness away from the traditional neurobiological approach*. Our research areas were: neurocardiology, solar and geomagnetic fluctuations and how it affects human autonomic nervous system, quantum physics of the human heart and brain and other related subjects. The heart as the dominant energetic organ of the human body and the role of heart rate variability (HRV) and its orchestrating symphony in the human body and the universe were the illuminators and the distinguished new scientific arena of the King of Organs Congresses.

5.1 Neurocardiology and the heart brain neurodynamics

The field of neurocardiology is relatively new discipline which was discussed first time in a scientific conference in King of Organs 2006,Saudi Arabia. The meticulous and sophisticated neurological afferent pathways (**Figure 4**) as well as energetic dominance of the heart over the brain was astonishing for the modern scientific communities. The amplitude of the cardiac electrical signal is about 60 times greater in amplitude compared to the brain while the electromagnetic field of the heart is approximately 5000 times stronger than the brain and can be detected six feet away from the body with sensitive magnetometers. Other ways the heart communicate the brain are hormonal and biophysical.

John and Beatrice Lacey during 1960s and 1970s created a massive drift in the modern psychophysiological research with their publications on human heart –brain communication [66, 67]. An important land mark in the field was there observations that afferent input from the heart and cardiovascular system could significantly affect perception, cognitive functions and behavior. This was neurophysiological evidence signifying that sensory and motor integration could be modified by cardiovascular activity. The heart behaves as if it had a mind of its own. In contradiction to Cannon theory of homeostasis, Laceys showed that patterns of physiological responses were affected as much by the context of a specific task and its requirements as by emotional stimuli. A phenomenon called by Laceys the *directional fractionation* denotes paradoxical heart rate response as it decelerated and blood pressure *decreased*, while simultaneously recorded parameters such as, respiratory rate, pupillary dilation and skin conductance all increased as expected. Later, cognitive performance fluctuated at a rhythm around 0.1 Hz was demonstrated by Velden and Wolk and showed that the modulation of cortical function was via the heart's influence was due to afferent inputs on the neurons in the thalamus, which globally synchronizes cortical activity and the consciousness phenomena [68, 69]. A critical observation here is the finding that "pattern and stability" (of the rhythm) of the heart's afferent inputs, rather than the number of neural bursts within the cardiac cycle that will modulate thalamic activity, which in turn has global effects on brain function and ultimately the consciousness experience [70].

DOI: http://dx.doi.org/10.5772/intechopen.95040

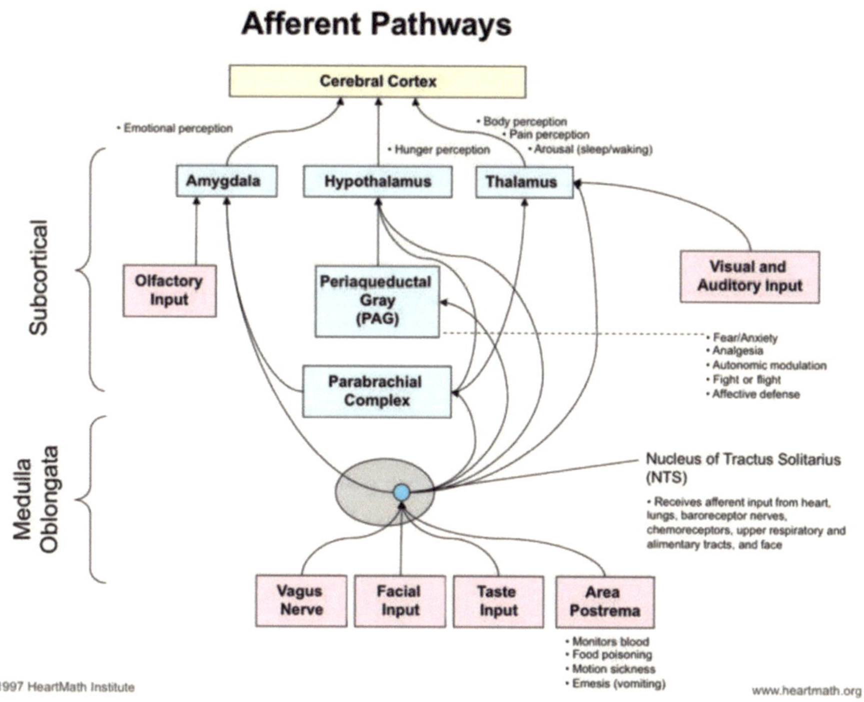

Figure 4.
The currently known afferent pathways by which information from the heart and cardiovascular system modulates brain activity. The nucleus of tractus solitarius (NTS) direct connection to the amygdala, hypothalamus and thalamus is shown. In addition there is emerging evidence of the presence of a pathway from the dorsal vagal complex that travels directly to the frontal cortex.

Growing body of respectful research has since been accumulating indicating that afferent information processed by the intrinsic cardiac nervous system can influence activity in the frontocortical areas and motor cortex, affecting psychological factors and the consciousness experience elements such as attention level, motivation, perceptual sensitivity, and emotional processing [70–72].

5.2 The revolutionary paradigm, the heart detect stimulus before the brain and brain neural events are locked to heartbeats

One of the strategic scientific, philosophical, as well as conceptual turning points that emanates from the basic science and neuroscientific arena is the accumulating evidence of the precedence of the heart detection of sensory stimulus before the brain. Hyeong-Dong Park in nature neuroscience, documented neural events locked to heartbeats before stimulus onset predict the detection of a faint visual grating in two regions that have multiple functional correlates and that belong to the same resting-state network:the posterior right inferior parietal lobule and the ventral anterior cingulate cortex **Figure 5** [73].

There is compelling evidence to suggest the physical heart is coupled to a field of information not bound by the classical limits of time and space [74]. Rigorous experimental study demonstrated the heart receives and processes information about a future event before the event actually happens. The study's results provide surprising data showing that both the heart and brain receive and respond to pre-stimulus information about a future event before it occurs but the heart proceeded the brain by 1.3 seconds which is truly too long time in the scale of neural impulse transmission which is counted with milliseconds (**Figure 6**).

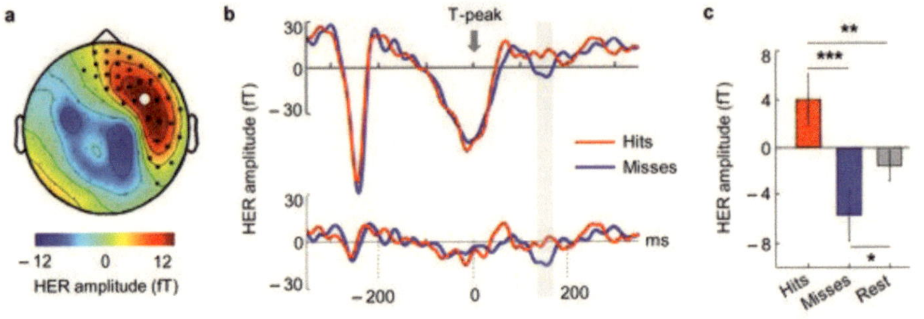

Figure 5.
Neural events locked to heartbeats before stimulus onset predict the conscious detection of a faint visual grating in the posterior right inferior parietal lobule and the ventral anterior cingulate cortex [73]. (HER): heartbeat-evoked response.

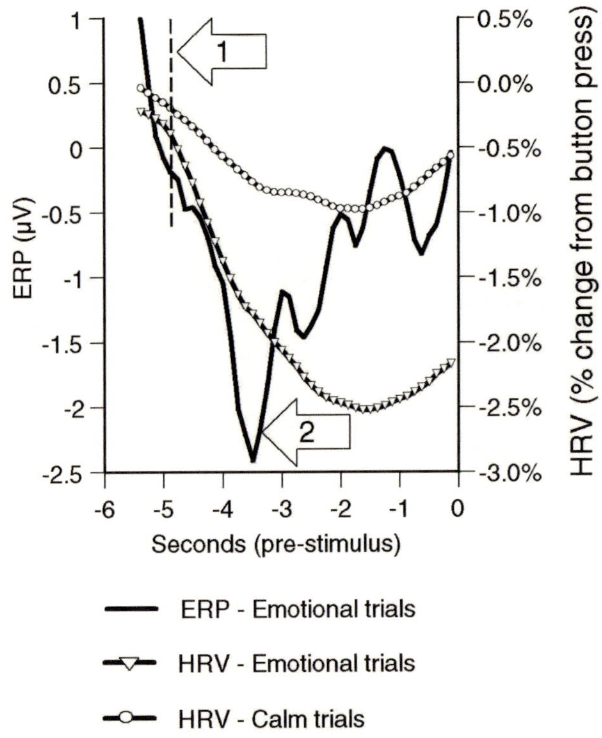

Figure 6.
Temporal dynamics of heart and brain pre-stimulus responses. Sharp downward shift about 4.8 seconds prior to the stimulus (arrow 1) is seen. The emotional trials ERP showed a sharp positive shift about 3.5 seconds prior to the stimulus (arrow 2). This positive shift in the ERP denotes the time the brain "knew" the nature of the future stimulus. The time difference between these two events suggests that the heart received the intuitive information about 1.3 seconds before the brain. Heartbeat-evoked potential analysis confirmed that a different afferent signal was sent by the heart to the brain during this period. (ERP) is event-related potential at EEG site FP2.HRV is heart rate variability [74].

5.3 *The vagal nerve is never vagal*, the afferent cardiac neuronal traffic and consciousness

Astonishing fact of the vagus nerve (means the nerve with unknown role) called sometimes, the tenth cranial nerve is the fact that it has very significant afferent neurons beside what we taught about its efferent neurons in our medical schools. 85–90% of the fibers in the vagus nerve are afferent [75]. The majority of higher brain centers, as well as emotional experience and cognitive processes are operated by Cardiovascular related afferent neural traffic [76]. Numerous brain centers including the thalamus, hypothalamus, and amygdala are connected to cardiovascular afferents. Range of frequencies of complex afferent information related to mechanical and chemical factors is continuously sent to the brain and is over time scales ranging from milliseconds to minutes [77]. Vagal afferent nerve stimulation causing increases traffic over the normal intrinsic levels in the thalamic pain pathways in the spinal cord will inhibit those pathways. In addition, vagal afferent nerve stimulation was shown to reduces migraine and cluster headaches and to improve cognitive processing and memory [78]. Activating afferent input with vagal nerve stimulation (VNS) is apparently heralding a new era in medical therapeutics as it proves effective in many psychophysiological disorders including epilepsy, obesity, depression, anxiety, autism, alcohol addiction, mood disorders, as well as multiple sclerosis, and traumatic brain injury [79, 80]. The cardiac coherence training is known to intensify afferent vagal neuronal inputs to cortical and subcortical systems and to Neuronal Correlates of Consciousness (NCC) with long term capabilities to reset the reference set up points resulting in increased afferent nerve activity noninvasively and ultimately improves psychophysiological parameters and the consciousness experience. There is thus a need to explore novel ways of repairing lost consciousness. Vagus nerve stimulation (VNS) may also contribute to breaking advances in awakening the unconscious vegetative state patient as approved by improvement in behavioral responsiveness and enhanced brain connectivity patterns. The vagus nerve carries afferent connections to the deep nuclei of the brain via the nucleus solitaries (see **Figure 4**). These afferent connections have multiple consciousness related targets, which include the thalamus, amygdala, reticular formation, hippocampus, raphe nucleus, and the locus coeruleus. VNS will create improved global neurostimulation state leading to promoted spread of cortical signals and caused an increase of metabolic activity leading to behavioral improvement as measured with the Coma Recovery Scale-Revised (CRS-R) scale [81]. Theta waves dominance were shown in the right inferior parietal and the parieto-temporal-occipital border, a region known to be instrumental in conscious awareness. *Improvement in long white matter tracts namely the corticocortical and thalamocortical disconnected pathways by vagus nerve stimulation is a true revolutionary therapeutic option. Today it is conspicuous to the whole medical communities that* **vagal nerve is never vagal.**

5.4 Cardiac coherence: repatterning psychophysiological neural networks and consciousness experience

McCraty and colleagues introduced the term physiological coherence to describe the degree of order, harmony, and stability in the various rhythmic activities within living systems over any given time period [82]. This harmonious order signifies a coherent system that has an efficient or optimal physiological functioning which will be reflected in more resilient personality and higher consciousness. Physiological coherence (also referred to as cardiac coherence) can be measured by HRV analysis where more ordered sine like HRV pattern will be seen around frequency of 0.1 Hz (10 seconds) which will be seen as very narrow, high-amplitude peak in the low frequency (LF) region of the HRV power spectrum with no major peaks in the VLF or HF regions [83].

Ground breaking discovery emphasizing the ability of afferent cardiac signals to reprogram the cortical and subcortical neural networks is what we describe as the *repatterning process* in the neural architecture, where coherence becomes established as a new, stable baseline reference memory [84]. Coherence is adaptable to be the new set point or the default reference point, facilitating the ability to self-regulate

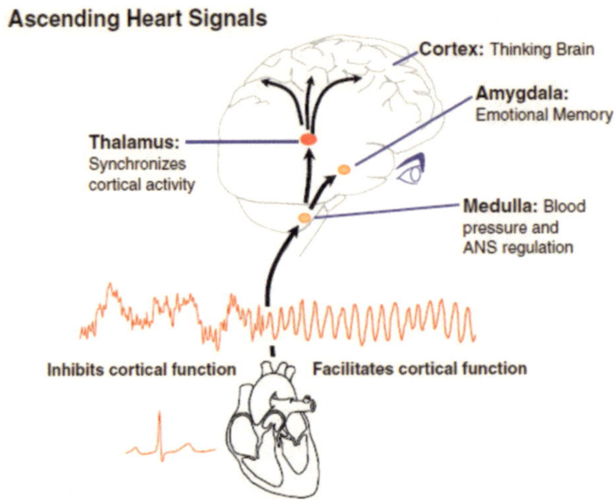

Figure 7.
Heart activity affects brain function. The ascending heart signals impact autonomic regulatory centers in the brain and cascade up to higher brain centers involved in emotional and cognitive processing, including the thalamus, amygdala, and cortex [83].

stress and emotions, a process that then becomes like a habit and eventually automatic [85–89]. Mental and emotional flexibility to remain in self-directed control is a well-known outcome with repeated coherence training. This will end up with more resilient personality and psychophysiological well being including capability to reduce systemic blood pressure without medications [70]. It also builds one's capacity to access intuitive state with higher consciousness to achieve intelligent life options easier through what we can call heart intelligence (**Figure 7**).

5.5 The heart signature on the brain interoception: heartbeat evoked potentials (HBEPs)

Heartbeat evoked potentials (HEPs) are segments of electroencephalogram (EEG) that are synchronized to the heartbeat. The ECG R-wave is used as a timing source for signal averaging, resulting in waveforms known as HEPs. Based on animal studies, Those cardiac afferents are transmitted to cortical areas including the insula, amygdala, somatosensory cortex and cingulate cortex, through subcortical relays such as the nucleus of the solitary tract, parabrachial nucleus, and thalamus Changes in these evoked potentials associated with the heart's afferent neurological input to the brain are detectable between 50 and 550 ms after each heartbeat [70].

Initiation of negative or positive emotion conditions by recalling past events reduced HRV and N250 amplitude. In contrast, resonance frequency breathing with HRV frequency around the 0.1 Hz peak increased HRV and HRV coherence above baseline and increased N250 amplitude [90]. We and others thought of HEPs as a neural marker of cardiac-related cortical processing in in consciousness and other diverse cognitive functions. Different afferent input mechanisms from the heart to the brain during different emotions and HRV can be identified using HEPs. Hyeong-Dong Park et al., found that neural responses to heartbeats can be recorded mainly in the insula (i.e., anterior, posterior) and operculum (i.e., frontal, central, posterior)., although it can be found in other regions distributed across the brain including the amygdala and fronto-temporal cortex [91]. It is known that insula is the primary cortical projection site of interoceptive signals. It is interesting to know that the HBEP is significantly higher during interoceptive compared to exteroceptive attention, in a time window of 524–620 ms after the R-peak [92].

5.6 Consciousness patterns, heart and brain interconnectivity

Similarities of basic frequencies, harmonics, magnetic field intensities, voltages, band widths, and energetic solutions between the Schumann resonances in the space between earth and ionosphere and the activity within the human cerebral cortices suggest the capacity for direct interaction [93]. Every cell in our body is bathed in an internal and external environment of fluctuating invisible magnetic forces that can affect virtually every cell and circuit in biological systems [94]. Therefore, it should not be surprising that numerous physiological rhythms in humans heart and brain and global collective behaviors are not only synchronized with solar and geomagnetic activity, but disruptions in these fields can create adverse effects on human health and behavior. The most sensitive body systems to those fluctuating electromagnetic environments are the heart and brain [95]. The heart is the largest dynamic organ in the human body. No surprise that the heart magnetic field is the strongest rhythmic field produced by the human body. The second strongest magnetic generator is the brain. The primary source of the electromagnetic activity of the brain measured from the scalp and emerges from the cerebral cortices because of the parallel arrangement of the dendrite-soma-axo orientations perpendicular to the surface for most of the approximately 20 billion neurons. Superimposed upon the steady potential are fluctuating voltages that define the electroencephalogram (EEG). It is not surprising that the heart's electrical field is about 60 times greater in amplitude than the electrical activity generated by the brain. There is a direct mathematical relationship between the HRV *patterns* and the spectral information encoded in the magnetic field [96]. The coherence model predicts that different emotions are reflected in state-specific *patterns* in the heart's rhythms regardless of the heart rate. *Patterns* in the activity of cardiovascular afferent neuronal traffic can significantly influence cognitive performance, emotional experience and self regulatory capacity via inputs to the thalamus, amygdala and other subcortical structures. There is 75% accuracy rate in detection of discrete emotional states from the HRV signal using a neural network approach for pattern recognition. It was found that information reflecting one's emotional state is encoded in the *patterns* of the HRV waveform and in addition, is contained in the heart's electromagnetic field radiated into the environment [96, 97]. When an individual is in a heart coherent state, the heart's magnetic field also has a more coherent structure. Information also is encoded in the interbeat intervals of the pressure and electromagnetic waves produced by the heart. It was shown that when two people are in a loving relationship that their hearts rhythms can synchronize even at great distances apart (Presented in King of Organs Congress 2019,permission from author Peter Granger). As the heart secretes a number of different hormones with each contraction, there is a hormonal pulse pattern that correlates with heart rhythms. In addition to the encoding of information in the space between nerve impulses and in the intervals between hormonal pulses, it is likely that information also is encoded in the interbeat intervals of the pressure and electromagnetic waves produced by the heart. This supports Pribram's proposal that low frequency oscillations generated by the heart and body in the form of afferent neural hormonal and electrical *patterns* are the carriers of emotional information and the higher frequency oscillations found in the EEG reflect the consciousness. The correlation of those physiological patterns to the Neuronal Correlates of Consciousness (NCC) and the correlation of this possible link to consciousness is to be investigated in the future.

5.7 Solar and planetary geomagnetic activity, the delicate orchestration of neuronal cardiac afferents to cerebral cortex

Resonance refers to vibration of large amplitude in electrical or mechanical system caused by a relatively small periodic stimulus of the same or nearly the same period as the natural vibration period of the system. The concept of resonance and its implementations in physiological as well as astrophysical rhythms is of critical significance for life on earth and to human consciousness experience. All biological systems on the planet are exposed to an external and internal environment of fluctuating invisible wide range of magnetic fields frequencies. These fields can affect virtually every cell and circuit to a greater or lesser degree. Numerous physiological rhythms have been shown to be synchronized with solar and geomagnetic activity. Geomagnetic and solar influences affect a wide range of human rhythmic systems with the nervous and cardiovascular systems, with their significant contribution to consciousness, being the most clearly impacted [70]. Sharp variations of sudden and sharp nature of geomagnetic, solar activity and its resultant geomagnetic storms can act as stressors, which has the capacity to alter body regulatory processes and rhythmic systems such as melatonin/serotonin balance, blood pressure, breathing, reproductive, immune, neurological, and cardiac system processes [98–101]. In the clinical arena significant increases in hospital admissions for depression, mental disorders psychiatric admission, homicides, suicide attempts, and traffic accidents are associated with planetary geomagnetic disturbances [102–108]. Increase incidence of myocardial infarctions, vascular variability disorders, local and global communication between humans during geomagnetic disturbances are all denotes that brain and cardiovascular systems are clear targets for the planetary geomagnetic disturbances [109–114]. Exacerbation of present disease like development of cardiac arrhythmias and epilepsy is well known during disturbed geomagnetic activity. Low frequency magnetic oscillations, around 3 Hz, was observed to cause Altered EEG rhythms with sedative effect [115]. Applying the lowest Schumann Resonance (SR) frequency of 7.8 Hz with 90 nano Tesla for 1.5 hours was found to be cardioprotective from stress conditions with reduction of the amount of CK released to the buffer, during normal conditions, hypoxic conditions and oxidative stress induced by 80 μM H2O2 [116]. The longest record in human history of human heart rate variability (HRV) synchronized with Solar Wind indices, Shumann Resonances (SR) and Galactic Cosmic Rays (GCR) monitoring was achieved by our group [117]. Schumann resonance frequency is 7.83 hertz (Hz), with a (day/night) variation of around ±0.5 Hz. The higher frequencies are ~14, 20, 26, 33, 39 and 45 Hz, all of which closely overlay with alpha (8–12 Hz), beta (12–30 Hz) and gamma (30–100 Hz) brain waves. The delicate orchestration of this universal symphony and vibrations with the human autonomic nervous system (ANS) that interacts with cerebral cortex and control heart rhythm, respiration, digestive functions and other involuntary activities was investigated. We were able to confirm that changes in solar and geomagnetic activity during periods of normal undisturbed activity affect daily ANS activity. In an other publication, we were able to document significant correlations between the group's HRV and solar wind speed, Kp, Ap, solar radio flux, cosmic ray counts, Schumann resonance power, and the total variations in the magnetic field [110] This affect is initiated at different times after the changes in the various environmental factors and persist over varying time periods. Peaks of increased solar activity occurs every 10.5 to 11 years. During those peaks, the sun emits increased ultraviolet (UV) energy and solar radio flux, which is measured by the 2.8 GHz signal (F10.7) [110] We considered Solar wind intensity as biological stressor as increase in its intensity is well correlated to increase heart rate. Galactic Cosmic Rays (GCR) are highly energetic particles that originate outside the solar system and are likely formed by nuclear explosive events in supernova and other mega giant galaxies. These highly energetic particles consist of fully ionized nuclei ranging from hydrogen, accounting for approximately 89% of the GCR spectrum, to trace amounts of uranium. The planetary magnetic field and the solar winds are protective for life on earth from this extremely ionized rays. We documented that human HRV with its modulatory effect on the consciousness pillars through ascending neuronal input to cortical and sub cortical structures increases with rise of the three major universal vibrations that we examined: Solar Winds, Shumann Resonances (SR), and the Galactic Cosmic Rays (GCR). This complex interaction between HRV and those environmental energetic fields may contribute to the human knowledge about the pathomechanistic effects on human psychphysiological homeostasis and the consciousness experience.

6. The realm of consciousness, from within the skull to the sky

6.1 Classical mechanics cannot explain consciousness

It is conspicuous from the previous sections that the neuronal firing of brain structures is not enough to explain subjective consciousness experience. Quantum physicists Larissa Brizhik and Emilio DelGiudice suggested that the most likely physical agent that can continuously provide an exchange of information between living systems within the larger ecosystem is the magnetic fields. According to the quantum field theory, potentials of the magnetic field, governs the dynamics of biological systems and the whole ecosystem. As a matter of fact, the planetary magnetic field is ubiquitous and involved in the deep behavior of biology. Animals can detect the Earth's magnetic field through magnetoreception-related photoreceptor cryptochromes [118] through which the planetary magnetic field guides the different species in their thousands of miles migration in land and oceans. The field causes the emergence of the coherent structures, which, in view of their coherence, openness and nonlinearity, are able to self-organize and form a chain of hierarchical levels of ecosystems [119] Coherence in the quantum language implies correlations, connectedness, consistency, efficient energy utilization, and the concept of global order, where the whole is greater than the sum of its individual parts. In medicine we refer to coherence to implies a harmonious relationship, correlations and connections between the various parts of a system. The Wight of evidence towards new evolutionary paradigm of the origin and effect of human consciousness with mutual effect to the environment is prevailing.

Evidence is accumulating supporting the hypothesis that our consciousness can even influences our physical world. Random number generators (RNGs) are one tool used to evaluate micro-psychokinesis or our ability to affect the physical world with our consciousness. Research conducted by the Global Consciousness Project (GCP) (which maintains a worldwide network of random number generators running constantly at about 60 locations around the world, sending streams of 200-bit trials generated each second to be archived as parallel random sequences), has found that human emotionality affects the randomness of these electronic devices in globally correlated manner. Roger Nelson who is the founder of GCP reported in a recent publication multiple examples of striking similarity between event-related brain potentials and event-related correlations in random data [119] (**Figure 8**). If all living systems are indeed interconnected and communicate with each other via biological and electromagnetic fields, it stands to reason that humans can work together in a co-creative relationship to consciously increase global coherence and raise the global consciousness. It is conspicuous that classical mechanics cannot explain consciousness. Quantum consciousness is the science that incorporate conceptual discussion of phenomenon of quantum mechanics like entanglement and superposition to explore the deep science of human consciousness.

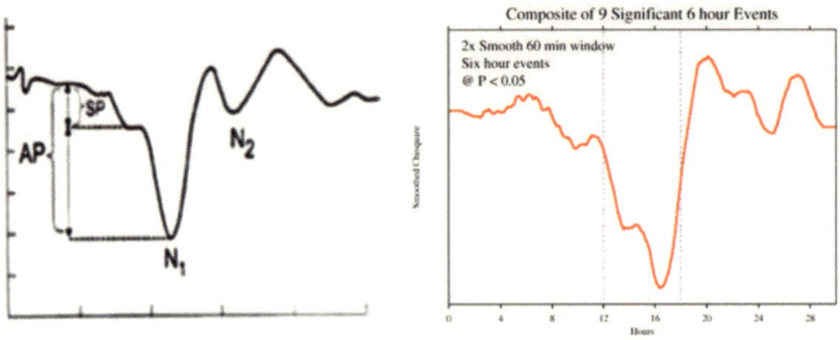

Figure 8.
Striking similarity between evoked potential (EP) from an auditory stimulus (the black) and composite of GCP data from nine 6 hour events (the red) [119].

6.2 The quantum consciousness

The idea that quantum mechanics has something to do with the workings of the mind was developed by Eugene Wigner, Hungarian-American theoretical physicist and Nobel Prize Laurete in Physics in 1963,who proposed that the wave function collapses due to its interaction with consciousness. Freeman Dyson argued that "mind, as manifested by the capacity to make choices, is to some extent inherent in every electron. David Bohm is theoretical physicists who contributed significantly to quantum theory, neuropsychology and the philosophy of mind. He stimulated new era of conceptual approach to consciousness with more fundamental level in the universe. He claimed both quantum theory and relativity pointed to this deeper theory, which he formulated as a quantum field theory. Bohm's proposed implicate order which applies both to matter and consciousness. He suggested that it could explain the relationship between them. Bohm's views mind and matter as projections into our explicate order from the underlying implicate order This more fundamental level was proposed to represent an undivided wholeness and an implicate order, from which arises the explicate order of the universe as we experience it. Holonomic brain theory is a branch of neuroscience investigating the idea that human consciousness is formed by quantum effects in or between brain cells. This specific theory of quantum consciousness was developed by neuroscientist Karl Pribram initially in collaboration with David Bohm. In addition to the neuroanatomical components of the human brain including the large fiber tracts in the brain, neurotransmissions also occurs in dendrites and other webs of fine fiber branche, that form webs. Due to the billions of action potentials and neural impulse formations, dynamic electrical fields will result around these dendritic trees. Those dendritic trees can affect other surrounding neurons without physical contact between them by entanglement. In this way, processing in the brain can occur in a non-localized manner. An energy-based concept of information was described by Dennis Gabor, who invented the hologram in 1947,which he described as quanta of information. Later on, he won Nobel prize in physics for this invention in 1971.Kal H.Pribram's holonomic model of brain processing was described in his 1991 *Brain and Perception* book which include his perspectives on human consciousness with David Bohm. It describes human cognition by modeling the brain as a holographic storage network. Pribram suggests these processes involve electric oscillations in the brain's dendritic networks, which are different from the more commonly known action potentials involving axons and synapses. These oscillations are waves and create wave interference patterns in which memory is encoded naturally. The waves are found to be analyzable by Fourier transform. Gabor, Pribram and others noted the similarities between these brain processes and the storage of information in a hologram. Pribram's holonomic model contributes significantly to human consciousness understanding, specially to the fast associative memory and the non-locality of memory. In 1991 Orchestrated Objective Reduction (Orch-OR) theory was introduced by physicist Roger Penrose and anesthesiologists Stuart Hameroff. It is a biological philosophy of mind that postulates that at the quantum level consciousness originates inside neurons, rather than the traditional perspective that it is a product of connections between neurons. The interpreting mechanism is contributed to non computational quantum process performed by quantum bits (qubits) formed collectively on cellular microtubules- called objective reduction. The qubits are based on oscillating dipoles(either electric or magnetic) forming superposed resonance rings in helical pathways throughout lattices of microtubules. Orchestration refers to the hypothetical process by which microtubule-associated proteins (MAPs) and other connective proteins, orchestrate qubit state reduction through modification of space time-separation of their superimposed states. Penrose was faced with a wave of criticism which is in our view not justified as his opponents were too limited with their perspective for human consciousness within the today computational intelligence language. In addition, postulating intuitive thoughts in the context of acceptable scientific language is well known and accepted approach in the philosophy of science. Other important contributors in the field are Hiroomi Umezawa and collaborators who proposed a quantum field theory of memory storage which is fundamentally different from the Penrose-Hameroff theory. In 1967, Hiroomi Umezawa together with L.M. Ricciardi, proposed a quantum theory of the brain which posits a spatially distributed charge formation exhibiting spontaneous breakdowns at micro levels as the basis for processing at macro levels. According to this model, the information resides in the virtual field associated with the dynamics of the cellular matter. Hiroomi Umezawa was known by his extreme originality. His approach was built upon by Karl Pribram and many others and expanded by Giuseppe Vitiello to a dissipative quantum model of brain. An other pioneer in the field is Henry Pierce Stapp. He is American mathematical physicist, known for his work in quantum mechanics who favors the idea that quantum wave functions collapse only when they interact with consciousness. According to Stapp hypothesis alternative quantum possibilities when exposed to conscious mind will select one. His explanation hypothesis differs from that of Penrose and Hameroff. Stapp postulates a process of global collapse through an effect on the synapses by exploitation certain aspects of quantum Zeno effect. David Pearce, British philosopher has conjectured that unitary conscious minds are physical states of quantum coherence (neuronal superpositions). It is clear at this point that scientist from different disciplines over the last 8 decades were trying to come closer to the absolute fact of consciousness. Each theory discussed has its strength and weaknesses but all lack the comprehensive universal perspective incorporating the origin of consciousness with more homogenous incorporation of current theories in a stronger model capable of bringing us closer to consciousness realm closer than any time ever.

7. Conclusion

In this chapter we investigate the elusive issue of human consciousness. We introduce revolutionary paradigm in the time line of consciousness science, where we discuss a comprehensive perspective of the process of consciousness of neurobiological and astrophysical bases. Our new perspective is built on our work confirming the symphony interplay of human ANS represented by HRV on one hand and Shumann Resonances, Solar Wind Indices and Cosmic Rays on the other hand. In addition to up to date discussion on the neuroanatomical aspects of consciousness, the delicate and powerful contribution of cardiac afferent input to brain consciousness related cortical and subcortical structures and pathways and heartbeat evoked potentials (HEP) is discussed. The role of the quantum principles and magnetic potentials in the universal information processing is emphasized. Our new perspective is complementary but never competitive to the quantum consciousness theories discussed especially the theories of Karl Primbram-David Bohem, Penrose and Stuart Hameroff, and Pierce Stapp. This new comprehensive understanding of human consciousness should bring many scientific disciplines closer to illustrate the necessity of the intelligent blend of science branches to solve historical human issues in medicine, science, philosophy, and religion.

Author details

Abdullah Abdulrhman Al Abdulgader
Senior Congenital Cardiologist and Invasive Electrophysiologist, Prince Sultan Cardiac Center, Alhasa, Hofuf, Saudi Arabia

*Address all correspondence to: kidsecho@yahoo.com

References

[1] Lewis CS. Studies in Words. Cambridge: Cambridge University Press; 1960.

[2] Bogen JE. On the neurophysiology of consciousness: An overview. Conscious Cogn 1995;4:52-62.

[3] Edelman GM. The remembered present: A biological theory of consciousness. New York:Basic Books; 1989.

[4] Newman J, Baars BJ. A neural attentional model of access to consciousness: A global workspace perspective. Cogn Neurosci 1993;4:255-90.

[5] Flohr H. An information processing theory of anaesthesia. Neuropsychologia 1995;33:1169-80.

[6] Sheinberg DL, Logothetis NK. The role of temporal cortical areas in perceptual organization.Proc Natl Acad Sci U S A 1997;94:3408-13.

[7] Crick F, Koch C. Are we aware of the neural activity in primary visual cortex. Nature 1995;375:121-3.

[8] Hirstein W. The contribution of prefrontal executive processes to creating a sense of self. Mens Sana Monogr 2011;9:150-8.

[9] Milner AD, Goodale MA. The Visual Brain in Action. Oxford: Oxford University Press; 1995.

[10] Aleksander I. How to build a Mind. In: Mapping the Mind Series. London: Weidenfeld and Nicolson; 2000.

[11] Moruzzi G, Magoun HW. Brain stem reticular formation and the activation of EEG. Electroencephalogr Clin Neurophysiol 1949;1:455-73.

[12] Tononi G, Koch C. The neural correlates of consciousness: An update. Ann N Y Acad Sci 2008;1124:239-61.

[13] Tononi G. Phi: A voyage from the brain to the soul. New York: Pantheon Books; 2012.

[14] Young GB, Pigott SE. Neurobiological basis of consciousness. Arch Neurol 1999;56:153-7.

[15] Maquet P. Functional neuroimaging of normal human sleep by positron emission tomography. J Sleep Res 2001;9:207-31.

[16] Alkire MT, Haier RJ, Fallon JH. Toward a unified theory of narcosis: Brain imaging evidence for a thalamocortical switch as the neurophysiologic basis of anesthetic induced unconsciousness. Conscious Cogn 2000;9:370-86.

[17] Clare Timbie and Helen Barbas. Pathways for Emotions: Specializations in the Amygdalar, Mediodorsal Thalamic, and Posterior Orbitofrontal Network. The Journal of Neuroscience, August 26, 2015 • 35(34):11976-11987

[18] Balleine BW, Killcross S. Parallel incentive processing: An integrated view of amygdala function. Trends Neurosci 2006;29:272-9.

[19] Kim SJ, Loucks RA, Palmer AL, Brown AC, Solomon KM, Marchante AN, et al. Structural and functional connections of the amygdala: From normal emotion to pathological anxiety. Brain Behav Res 2011;223:403-10.

[20] Schmahmann JD. Disorders of the cerebellum: Ataxia, dysmetria of thought, and the cerebellar cognitive affective syndrome. J Neuropsychiatry Clin Neurosci 2004;16:367-78.

[21] 193.Schmahmann JD, Weilburg JB, Sherman JC. The neuropsychiatry of the cerebellum insights form the clinic. Cerebellum 2007;6:254-67.

[22] Strata P, Scelfo B, Sacchetti B. Involvement of cerebellum in emotional behavior. Physiol Res 2011;60:S39 S48.

[23] Schmahmann JD. In: The cerebellum and cognition. San Diego CA: Academic Press; 1997. p. 613-34.

[24] Hal Blumenfeld. Neuroanatomical Basis of Consciousness chapter. Book:The Neurology of Conciousness. December 2016.DOI: 10.1016/B978-0-12-800948-2.00001-7

[25] McCormick, D.A., 2002. Cortical and subcortical generators of normal and abnormal rhythmicity. Int. Rev. Neurobiol. 49, 99_114.

[26] Crick F, Koch C. Are we aware of the neural activity in primary visual cortex. Nature 1995;375:121-3.

[27] Rees G, Kreiman G, Koch C. Neural correlates of consciousness in humans. Nat Rev Neurosci 2002;3:261-70.

[28] De Sousa A. Towards An Integrative Theory Of Consciousness: Part 1 (Neurobiological And Cognitive Models). Mens Sana Monogr 2013;11:100-50.

[29] Michel M, Morales J. Minority reports: Consciousness and the Prefrontal cortex. Mind & Language. 2019;1-21. https://doi.org/10.1111/mila.12264

[30] Andrea E. Cavanna1, and Michael R. Trimble. The precuneus: a review of its functional anatomy and behavioural correlates. Brain (2006), 129, 564-583.

[31] Steriade, M., 2004. Acetylcholine systems and rhythmic activities during the waking—sleep cycle. Prog. Brain Res. 145, 179_196.

[32] Shi YF, Yu YQ. *Zhejiang Da Xue Xue Bao Yi Xue Ban*. 2013;42(5):583-590.

[33] Rasmusson, D.D., Clow, K., Szerb, J.C., 1994. Modification of neocortical acetylcholine release and electroencephalogram desynchronization due to brainstem stimulation by drugs applied to the basal forebrain. Neuroscience. 60 (3), 665_677.

[34] Dringenberg, H.C., Olmstead, M.C., 2003. Integrated contributions of basal forebrain and thalamus to neocortical activation elicited by pedunculopontine tegmental stimulation in urethane anesthetized rats. Neuroscience. 119 (3), 839_853.

[35] Blanco-Centurion, C.A., Shiromani, A., Winston, E., Shiromani, P.J., 2006. Effects of hypocretin-1 in 192-IgG-saporin-lesioned rats. Eur. J. Neurosci. 24 (7), 2084_2088.

[36] Freund, T.F., Meskenaite, V., 1992. gamma-Aminobutyric acidcontaining basal forebrain neurons innervate inhibitory interneurons in the neocortex. Proc. Natl. Acad. Sci. USA. 89 (2), 738_742.

[37] Saper, C.B., Fuller, P.M., Pedersen, N.P., Lu, J., Scammell, T.E., 2010. Sleep state switching. Neuron. 68 (6), 1023_1042.

[38] Mesulam, M.M., Mufson, E.J., 1984. Neural inputs into the nucleus basalis of the substantia innominata (Ch4) in the rhesus monkey. Brain. 107 (Pt 1), 253_274.

[39] Parent, A., Steriade, M., 1984. Midbrain tegmental projections of nucleus reticularis thalami of cat and monkey: a retrograde transport and antidromic invasion study. J. Comp. Neurol. 229 (4), 548_558.

[40] Brown, E.N., Lydic, R., Schiff, N.D., 2010. General anesthesia, sleep, and coma. N. Engl. J. Med. 363 (27), 2638_2650.

[41] Foote, S.L., Bloom, F.E., Aston-Jones, G., 1983. Nucleus locus ceruleus:

new evidence of anatomical and physiological specificity. Physiol. Rev. 63 (3), 844_914.

[42] Jacobs, B.L., Azmitia, E.C., 1992. Structure and function of the brain serotonin system. Physiol. Rev. 72 (1), 165_229.

[43] Hannon, J., Hoyer, D., 2008. Molecular biology of 5-HT receptors. In: Monti, J.M., Pandi-Perumal, S.R., Jacobs, B.L., Nutt, D.J. (Eds.)

[44] Dugovic, C., Wauquier, A., Leysen, J.E., Marrannes, R., Janssen, P.A., 1989. Functional role of 5-HT2 receptors in the regulation of sleep and wakefulness in the rat. Psychopharmacology (Berl). 97 (4), 436_442.

[45] Sowers, L.P., Massey, C.A., Gehlbach, B.K., Granner, M.A., Richerson, G.B., 2013. Suden unexpected death in epilepsy: fatal post-ictal respiratory and arousal mechanisms. Respir. Physiol. Neurobiol. 189 (2), 315_323.

[46] Bandyopadhyay, S., Hablitz, J.J., 2007. Dopaminergic modulation of local network activity in rat prefrontal cortex. J. Neurophysiol. 97 (6), 4120_4128.

[47] Combarros, O., Infante, J., Berciano, J., 2000. Akinetic mutism from frontal lobe damage responding to levodopa. J. Neurol. 247 (7), 568_569.

[48] Brown, R.E., Stevens, D.R., Haas, H.L., 2001. The physiology of brain histamine. Prog. Neurobiol. 63 (6), 637_672.

[49] Dringenberg, H.C., Kuo, M.C., 2003. Histaminergic facilitation of electrocorticographic activation: role of basal forebrain, thalamus, and neocortex. Eur. J. Neurosci. 18 (8), 2285_2291.

[50] McCormick, D.A., Williamson, A., 1991. Modulation of neuronal firing mode in cat and guinea pig LGNd by histamine: possible cellular mechanisms of histaminergic control of arousal. J. Neurosci. 11 (10), 3188_3199.

[51] Sakurai T, Amemiya A, Ishii M, Matsuzaki I, Chemelli RM, Tanaka H, Williams SC, Richardson JA, Kozlowski GP, Wilson S, Arch JR, Buckingham RE, Haynes AC, Carr SA, Annan RS, McNulty DE, Liu WS, Terrett JA, Elshourbagy NA, Bergsma DJ, Yanagisawa M (February 1998). "Orexins and orexin receptors: a family of hypothalamic neuropeptides and G protein-coupled receptors that regulate feeding behavior". Cell. **92** (4): 573-85. doi:10.1016/S0092-8674(00)80949-6

[52] Lazarus M, Oishi Y, Bjorness TE and Greene RW (2019) Gating and the Need for Sleep: Dissociable Effects of Adenosine A_1 and A_{2A} Receptors. *Front. Neurosci*. 13:740. doi: 10.3389/fnins.2019.00740

[53] Nitsch R. 2012. Gehirn, Geist und Bedeutung: Zur Stellung der Neurowissenschaften in der Leib-Seele-Diskussion. Münster (Germany): Mentis. p. 17.

[54] Nitsch R. 2012. Gehirn, Geist und Bedeutung: Zur Stellung der Neurowissenschaften in der Leib-Seele-Diskussion. Münster (Germany): Mentis. p. 17.

[55] Abdullah Alabdulgader. The Ancient Wisdom at Intersection with Modern Cardiac Sciences.Special issue.submitterd to Heart and Mind Journal

[56] Penfield W. 1938. The cerebral cortex in man: I. The cerebral cortex and consciousness. Arch Neurol Psychiatry. 40(3): 417-442.

[57] Kolb B, Teskey GC. 2012. Age, experience, injury, and the changing brain. Dev Psychobiol. 54(3):311-325.

[58] R. Nitsch1,2 and F. W. Stahnisch3,4. Neuronal Mechanisms Recording the Stream of Consciousness–A Reappraisal of Wilder Penfield's (1891-1976) Concept of Experiential Phenomena Elicited by Electrical Stimulation of the Human Cortex. Cerebral Cortex, September 2018;28: 3347-3355. doi: 10.1093/cercor/bhy085

[59] Bjorn Merker. Consciousness without a cerebral cortex: a challenge for neuroscience and medicine. Review Behav Brain Sci 2007 Feb;30(1):63-81; discussion 81-134. doi: 10.1017/S0140525X07000891.

[60] Penfield W. & Jasper, H. (1954) *Epilepsy and the Functional Anatomy of the Human Brain*, Boston, MA: Little and Brown.

[61] James M. Sprague. Interaction of Cortex and Superior Colliculus in Mediation of Visually Guided Behavior in the Cat. *Science* 23 Sep 1966: Vol. 153, Issue 3743, pp. 1544-1547. DOI: 10.1126/science.153.3743.1544

[62] Bjorn Merker "Consciousness without a cerebral cortex: A challenge for neuroscience and medicine" *Behavioral and Brain Sciences* **30**:63-81 (2007).

[63] Lionel Feuillet, Henry Dufour, Jean Pelletier. Brain of a white-collar worker. Lancet 2007; 370: 262

[64] Witelson SF, Kigar DL, Harvey T. The exceptional brain of Albert EInstein. *Lancet* 1999; **353**: 2149-53.

[65] *M Galaburda*. Albert Einstein's brain. THE LANCET • Vol 354 • November 20, 1999

[66] Lacey JI. Psychophysiological approaches to the evaluation of psychotherapeutic process and outcome. In: Rubinstein E, Parloff M, eds. *Research in Psychotherapy* Washington, DC: American Psychological Association, 1959:160-208.

[67] Lacey JI, Kagan J, Lacey BC, Moss HA. The visceral level: Situational determinants and behavioral correlates of autonomic response patterns. In: Knapp PH, ed. *Expression of the Emotions in Man*. New York: International Universities Press, 1963: 161-196.

[68] Wolk C, Velden M (1989) Revision of the baroreceptor hypothesis on the basis of the new cardiac cycle effect. In Psychobiology: issues and applications. Amsterdam: Elsevier Science Publishers B.V. 371-379.

[69] Lane RD, Reiman EM, Ahem GL, Thayer JF (2001) Activity in medial prefrontal cortex correlates with vagal component of heart rate variability during emotion. *Brain Cognit* 47: 97-100.

[70] Abdullah A Alabdulgader. The human heart rate variability; Neurobiology of psychophysiological well being and planetary resonance. Editorial. *Gen Int Med Clin Innov,* 2017Volume 2(2): 2-4. doi: 10.15761/GIMCI.1000141

[71] McCraty R, Atkinson M, Tomasino D, Bradley RT (2009) The coherent heart: heartbrain interactions, psychophysiological coherence, and the emergence of system-wide order. Boulder Creek, CA: Institute of Heartmath.

[72] Lane RD, Reiman EM, Ahem GL, Thayer JF (2001) Activity in medial prefrontal cortex correlates with vagal component of heart rate variability during emotion. *Brain Cognit* 47: 97-100.

[73] Hyeong-Dong Park, Stéphanie Correia, Antoine Ducorps, Catherine Tallon-Baudry (2014) Spontaneous fluctuations in neural responses to

heartbeats predict visual detection. *Nature Neuroscience* 17: 612-618.

[74] McCraty, R., M. Atkinson, and R.T. Bradley, *Electrophysiological evidence of intuition: Part 2. A system-wide process?* Journal of Alternative and Complementary Medicine, 2004. **10**(2): p. 325-336.(Plenary lecture in King of Organs 2008 Conference)

[75] Cameron,O.G.(2002). *Visceral Sensory Neuroscience:Interception*. New York: Oxford University Press

[76] McCraty,R.,Atkinson,M.,Toma sino,D.,andBradley,R.T.(2009b).The coherentheart:heart-brain interactions, psychophysiological coherence, and the emergence of system wide order. *Integral. Rev.* 5, 10-115.

[77] Armour,J.A.,andKember,G.C.(2004)."Cardiacsensoryneurons,"in *Basicand Clinical Neurocardiology*, edsJ.A.Arm ourandJ.L.Ardell(NewYork:Oxford UniversityPress),79-117.

[78] Hassert,D.L.,Miyashita,T.,andWillia ms,C.L.(2004).The effects of peripheral vagal nerve stimulation at memory-modulating intensity on norepinephrine out put in the basolateralamygdala. *Behav.Neurosci.* 118, 79-88.doi: 10.1037/0735-7044.118.1.79

[79] Kosel,M.,andSchlaepfer,T.E.(2003). Beyond the treatment of epilepsy: new applications of vagus nerve stimulation in psychiatry. *CNS Spectr.* 8, 515-521.

[80] Groves,D.A.,andBrown,V.J.(2005). Vagal nerve stimulation: are view of its applications and potential mechanisms that mediate its clinical effects. *Neurosci. Biobehav.Rev.* 29, 493-500.doi:10.1016/j.neubiorev.2005.01.004

[81] Tiller WA1, McCraty R, Atkinson M (1996) Cardiac coherence: a new, noninvasive measure of autonomic nervous system order. *Altern Ther Health Med* 2: 52-65. [Crossref]

[82] Rollin McCraty, Fred Shaffer (2015) Heart Rate Variability: New Perspectives on Physiological Mechanisms, Assessment of Self-regulatory Capacity, and Health Risk. *Glob Adv Health Med* 4: 46-61. [Crossref]

[83] Bradley, R.T., McCraty, R., Atkinson, M., Tomasino., D., *Emotion Self-Regulation, Psychophysiological Coherence, and Test Anxiety: Results from an Experiment Using Electrophysiological Measures.* Applied Psychophysiology and Biofeedback, 2010. **35**(4): p. 261-283.

[84] McCraty, R., et al., *The coherent heart: Heart-brain interactions, psychophysiological coherence, and the emergence of system-wide order*. 2006, Boulder Creek, CA: HeartMath Research Center, Institute of HeartMath, Publication No. 06-022.

[85] McCraty, R. and D. Childre, *The grateful heart: The psychophysiology of appreciation*, in *The Psychology of Gratitude*, R.A. Emmons and M.E. McCullough, Editors. 2004, Oxford University Press: New York. p. 230-255.

[86] McCraty, R., M. Atkinson, and D. Tomasino, *Impact of a workplace stress reduction program on blood pressure and emotional health in hypertensive employees.* Journal of Alternative and Complementary Medicine, 2003. **9**(3): p. 355-369.

[87] Martina Corazzol1, Guillaume Lio1, Arthur Lefevre,et al. Restoring consciousness with vagus nerve stimulation. Correspondence. Current Biology *27*, R979–R1001, September 25, 2017

[88] Alabdulgader, A., *Coherence: A Novel Nonpharmacological Modality for Lowering Blood Pressure in Hypertensive Patients.* Global Advances in Health and Medicne, 2012. **1**(2): p. 54-62.

[89] Abdullah Alabdulgader (2016) Modulation of heart rate variability: A novel nonpharmacological modality

for lowering blood pressure in in hypertensive patients. *J Clin Exp Cardiolog*

[90] Starr MacKinnon, Richard Gevirtz, Rollin McCraty, Milton Brown (2013) Utilizing Heartbeat Evoked Potentials to Identify Cardiac Regulation of Vagal Afferents During Emotion and Resonant Breathing. Appl Psychophysiol Biofeedback 38: 241-255. [Crossref]

[91] Hyeong-Dong Park1, Fosco Bernasconi1, Roy Salomon2, Catherine Tallon,et al. Neural Sources and Underlying Mechanisms of Neural Responses to Heartbeats, and their Role in Bodily Self-consciousness: An Intracranial EEG Study. Cerebral Cortex, July 2018;28: 2351-2364. doi: 10.1093/cercor/bhx136

[92] Frederike H.Petzschner et al, Focus of attention modulates the heartbeat evoked potential. NeuroImage Volume 186, 1 February 2019, Pages 595-606

[93] Kevin S. Saroka, Michael A. Persinger. Quantitative Evidence for Direct Effects Between Earth-Ionosphere Schumann Resonances and Human Cerebral Cortical Activity. *International Letters of Chemistry, Physics and Astronomy Online: 2014-10-02 ISSN: 2299-3843, Vol. 39, pp 166-194.doi:10.18052/www.scipress.com/ILCPA.39.166.2014 SciPress Ltd, Switzerland*

[94] *Rollin McCraty. SCIENCE OF THE HEART Exploring the Role of the Heart in Human PerformanceVolume 2. Published by: HeartMath Institute.*

[95] Halberg, F., et al., Cross-spectrally coherent ~10.5- and 21-year biological and physical cycles, magnetic storms and myocardial infarctions. Neuroendocrinology, 2000. **21**: p. 233-258.

[96] *McCraty R, Atkinson M, Tomasino D, Bradley R. The coherent heart: heartbrain interactions, psychophysiological coherence, and the emergence of system-wide order. Integr Rev (2009) 5(2):10-115.*

[97] *McCraty R. The energetic heart: bioelectromagnetic communication within and between people. In: Rosch PJ, Markov MS, editors. Bioelectromagnetic Medicine. New York: Marcel Dekker (2004). p. 541-62.*

[98] *Cherry, N. Schumann Resonances, a plausible biophysical mechanism for the human health effects of Solar/ Geomagnetic Activity.Natural Hazards* **26**, *279-331 (2002).*

[99] Ghione, S., Mazzasalma, L., Del Seppia, C. & Papi, F. Do geomagnetic disturbances of solar origin affect arterial blood pressure?Journal of Human Hypertension **12**, 749-754 (1998).

[100] *Hamer, J. R. Biological entrainment of the human brain by low frequency radiation. Northrop Space Labs, 65-199 (1965).*

[101] *Chernouss, S., Vinogradov, A. & Vlassova, E. Geophysical Hazard for Human Health in the Circumpolar Auroral Belt: Evidence of a Relationship between Heart Rate Variation and Electromagnetic Disturbances. Natural hazards* **23**, *121-135 (2001).*

[102] *Gordon, C. & Berk, M. The effect of geomagnetic storms on suicide. South African Psychiat Rev* **6**, *24-27 (2003).*

[103] *Kay, R. W. Geomagnetic Storms: Association with Incidence of Depression as Measured by Hospital Admission. British Journal of Psychiatry* **164**, *403-409 (1994).*

[104] *Kay, R. W. Schizophrenia and season of birth: relationship to geomagnetic storms. Schiz Res* **66**, *7-20 (2004).*

[105] *Nikolaev, Y. S., Rudakov, Y. Y., Mansurov, S. M. and Mansurova, L. G. Interplanetary magnetic field sector structure and disturbances of the central*

nervous system activity. Reprint N 17a, Acad. Sci USSR, IZMIRAN, Moscow, 29 (1976).

[106] *Oraevskii, V. N. et al. Effect of geomagnetic activity on the functional status of the body. Biofizika* **43**, *819-826 (1998).*

[107] *Halberg, F., Cornelissen, G., Panksepp, J., Otsuka, K. & Johnson, D. Chronomics of autism and suicide. Biomed Pharmacother* **59***(1), S100-108 (2005).*

[108] *Berk, M., Dodd, S. & Henry, M. Do ambient electromagnetic fields affect behaviour? A demonstration of the relationship between geomagnetic storm activity and suicide. Bioelectromagnetics* **27**, *151-155 (2006).*

[109] *Timofejeva I, McCraty R, Atkinson M, Joffe R, Vainoras A, Alabdulgader AA, Ragulskis M. Identification of a Group's Physiological Synchronization with Earth's Magnetic Field. Int J Environ Res Public Health. 2017 Sep 1;14(9):998. doi: 10.3390/ijerph14090998. PMID: 28862697; PMCID: PMC5615535.*

[110] McCraty R, Atkinson M, Stolc V, Alabdulgader AA, Vainoras A, Ragulskis M. Synchronization of Human Autonomic Nervous System Rhythms with Geomagnetic Activity in Human Subjects. Int J Environ Res Public Health. 2017;14(7):770. Published 2017 Jul 13. doi:10.3390/ijerph14070770

[111] Abdullah Alabdulgade, Rollin McCraty, Mike Atkinson, Alfonsas Vainoras, Kristina Berškiene, et al. (2015) Human Heart Rhythm Sensitivity to Earth Local Magnetic Field Fluctuations. Journal of Vibroengineering:17.

[112] Franz Halberg, Germaine Cornélissen, Rollin McCraty, Jerzy Czaplicki, Abdullah A. Al-Abdulgader (2011) Time Structures (Chronomes) of the Blood Circulation, Populations' Health, Human Affairs and Space Weather. World Heart Journal 3.

[113] Abdullah A. Al-Abdulgader, Germaine Cornélissen Guillaume, Franz Halberg (2011) Vascular Variability Disorders in the Middle East:Case Reports. World Heart Journal 2.

[114] Inga Timofejeva , Rollin McCraty, Mike Atkinson, Roza Joffe, Abdullah A. Alabdulgader , Alfonsas Vainoras, Mantas Landauskas, , Minvydas RagulskisGlobal Study of Human Heart Rhythm Synchronization with the Earth's Time Varying Magnetic Field,Underr review in the BMC Bioinformatics journal.

[115] Belov, D. R., Kanunikov, I. E. & Kiselev, B. V. Dependence of human EEG synchronization on the geomagnetic activity on the day of experiment. Ross Fiziol. Zh Im I M Sechenova **84**, 761-774 (1998).

[116] Elhalel G, Price C, Fixler D, Shainberg A. Cardioprotection from stress conditions by weak magnetic fields in the Schumann resonance band. Nature Scientific Reports, 2019;**9**(1)

[117] Alabdulgader A, McCraty R, Atkinson M, Dobyns Y, Vainoras A, Ragulskis M, et al. Long-Term Study of Heart Rate Variability Responses to Changes in the Solar and Geomagnetic Environment [Internet]. Nature News. Nature Publishing Group. 2018. Available from: https://www.nature.com/articles/s. 41598-018-20932-x

[118] Siying Qin1, Hang Yin1, Celi Yang, Yunfeng Dou, et al .A magnetic protein biocompass. NATURE MATERIALS. 16 NOVEMBER 2015 | DOI: 10.1038/NMAT4484

[119] Roger D. Nelson. The Global Consciousness Project's Event-Related Responses Look Like Brain EEG Event-Related Potentials.Journal of Scientific Exploration, Vol. 34, No. 2 pp. 246-267, 2020

Chapter 2

Dynamics of Praxis Functions in the Context of Maturation of the Parietal and Frontal Brain Regions in the Period 4-6 Years of Age

Neli Cvetanova Vasileva and Jivko Dimitrov Jekov

Abstract

In recent years, child neuropsychology has paid special attention to ontogenesis and trends in the development of practical functions during the preschool period, given their relationship to practical skills and children's readiness to learn. On the other hand, the dynamics of complex types of praxis is an indicator of the integration between the brain regions responsible for the perception, programming and recoding of motor patterns. The article presents a comparative analysis of data from a study of two types of praxis functions (dynamic praxis and spatial postural praxis) in children with typical development in the period 4–6 years. The specificity of the performance of neuropsychological tests is an indicator of the functioning and the degree of neuronal connectivity of the parietal and premotor regions of the left hemisphere. The data from the study show a similar trend in the dynamics of the studied functions and the influence on them of three independent factors: age, social conditions (type of settlement) and gender. Significant improvement in the performance of the tasks is observed in children at the age of 6, which is a reason to consider this age as critical for the maturation and neurophysiological connectivity of the structures of the parietal and premotor regions. The assessment of complex types of praxis in this period is an objective indicator of the neuropsychological development of children and has an indisputable prognostic effect for future learning disorders.

Keywords: child neuropsychology, dynamic praxis, spatial postural praxis, children with typical development, parietal and premotor regions, neuropsychological development, learning disorders

1. Introduction

In recent years, the attention of specialists is increasingly focused on the assessment of motor functions in the preschool period. One of the reasons is the growing number of children with delayed motor development, whose symptoms can be either leading (Developmental dyspraxia, Developmental Coordination Disorder) or part of other neurodevelopmental syndromes (Autism Spectrum Disorders, Developmental Dyslexia and Attention Deficit Hyperactivity Disorder). In any case, the incomplete formation of motor skills is accompanied by cognitive, language, and emotional disorders that have a negative impact on children's school readiness.

The objective analysis of the observed deficits is directly related to the differential diagnostic and prognostic aspects of child development. The latter are the subject of child neuropsychology, whose methodological tools are aimed at analyzing the formation of higher mental functions (gnosis, praxis and language) and their relationship with the maturation of different brain regions. The complex neurophysiological organization of these functions and the individual rates of development of the child's brain are a prerequisite for separating the neuropsychology of individual differences (differential neuropsychology). Developments in the field outline the natural stages and patterns of formation of higher mental functions, their sensitive periods and age norms.

The range of age norms, which determines the registration of developmental disorders, is related to the tendency to "go" beyond the traditional framework of pathology and draws attention to the stages and patterns of typical ontogenesis. The diagnosis of any mental function is based on the notions of its normative meanings and is important for identifying so-called "soft" developmental abnormalities [1]. Along with the general characteristics of the functions, the researchers' interest is focused on the variability and peculiarities of neuropsychological development, referred to as the "typology of the norm". This explains the increasing emphasis on the cases of the "low" child norm, defined as a risk for the development of specific learning difficulties [2].

The active inclusion of neuropsychological methods in the study of the child population is associated with new trends in the analysis of mental development - from purely diagnostic to prognostic; from finding isolated deficits to describing syndromes and developing adequate treatment strategies [3]. The changes also reflect the idea of replacing the static approach with a dynamic one, in which the analysis focuses on the interaction between brain structures and mental functioning in the context of social conditions [4].

Chronology and normative diversity in the development of praxis functions are one of the least developed units in child neuropsychology. These functions have a complex brain organization, including processes of spatial orientation, coordination, programming and recoding of motor models, which is why their assessment has important prognostic value for child development. The fact is that, unlike established tools for language and cognitive functions, the diagnosis of motor development has not yet developed a gold standard assessment tool [5]. One of the explanations is that the early developments on the problems of motor development are mainly in the field of psychology and refer to the first half of the 20th century. By the 1960s, the subject of research has shifted from the biology of children's motor behavior in the direction of language and cognitive development as genetically related to learning [6].

Scientific developments in recent decades are an example of compensating for this discrepancy and show increased interest in the laws of motor ontogenesis and its neurophysiological organization. This is largely due to the recognition that the level of motor development is a determining factor for growth and behavior [7]. This is a reason to assume that the identification of deficits in complex motor (practical) functions during the preschool period allows timely support and optimization of cognitive and emotional-behavioral development of children.

2. Brain organization of motor development

2.1 Conceptual foundations of development

Gabbard and co-authors [7] consider motor development as a change in motor behavior influenced by the interaction of biological factors and the influence of the environment (training and education). Discussions on the subject correspond to the theory of dynamic systems, according to which man is a dynamic and self-organizing unit, consisting of a large number of systems (nervous, muscular, cognitive, etc.), each of which has different levels of organization. From this position, development is seen as a process of constant change in behavioral patterns under the influence of the environment and tasks. The theory outlines three main variables - individual, task and environment, the interaction between which generates spontaneous adaptive behavior. According to the cited authors, dynamic systems should be considered as part of the global (general) development systems, which approach gives a broader perspective in the study and understanding of development. Attempts are also made to integrate the theory of dynamic systems in the process of motor therapy in some forms of pathology - Autism Spectrum Disorders, Developmental Coordination Disorder, post-stroke conditions [6].

By accepting the stimulating role of the environment as leading to development, the theory of dynamic systems differs significantly from the older neuronal theory of maturation, which emphasizes the role of the nervous system [8]. According to the neuronal maturation theory, the stimulating effect of the environment is limited by the genetically set stages of maturation of the nervous system. From this point of view, training and therapy can lead to changes in the development of a function only after the associated nerve structures reach a certain degree of maturity.

A compromise between the first two is the Neuronal Group Selection Theory, which views development as the result of a complex intertwining of information from genes and environment. It defines variability as a basic principle of typical ontogenesis, relating to all its parameters - duration of stages of development, motor, language and cognitive skills. Neuronal Group Selection Theory postulates the initial existence of complex vertical connections between a huge number of neurons at the cortical and subcortical level, united in dynamically changing network systems (neuronal groups) with the character of functional units. The structural and functional organization of neuronal groups varies and is selected depending on the stage of development, afferent information and behavioral requirements (in [8]).

Similar ideas are shared by the theory of neural modular organization of the central nervous system [9, 10], which considers the structural development of the cortex as related to the formation of neural ensembles (neural centers) underlying mental ontogenesis. Those of them, which have the same type of functions, are provided in a larger structure - modules, with the nature of the basic units for information processing. Neural ensembles in all parts of the cortex are in a process of constant change. Although they obey a single mechanism, they have uneven dynamics over time. The rate of brain transformation is heterochronous, and developmental changes are faster the smaller the child.

The last two theories correspond closely with the leading principle in child neuropsychology for heterochronous formation of higher mental functions [11–13]. According to him, the functional organization of mental development is subject to a certain chronological sequence, in which each function is distinguished by its chronological formula and cycle of development, specifically related to the stimuli of the environment. The uneven formation of the functions explains the differences in their sensitive periods and the anticipatory development of some of them. The combination of genetically determined heterochrony and environmental influences determine the variety of individual (phenotypic) variants of development, often located at both extremes of the age norm - high and low [14]. Phenotypic diversity is among the leading goals of differential neuropsychology, related to the analysis of variants in the formation of cortical–subcortical brain systems and partial retardation in the development of individual higher functions (gnostic, practical, linguistic) within the typical development [11].

The syndrome analysis has a direct connection with the regularities in the development, the main task of which is the assessment of the individual neuropsychic profile. Except in cases of neurodevelopmental disorders, it also applies to the variety of cases within the typical child development. In this regard, some authors [13] use the term "positive developmental syndrome" as a combination of functions that have reached a certain level of development (positive symptoms). Due to the rapid changes in cerebral functional systems in early childhood, the derivation of regulatory trends should cover close age periods.

At the same time, the objective assessment of each individual case requires consideration of the dynamics of developmental changes related to the analysis of the beginning and direction of the developmental trajectory of the specific phenotype [15]. It should be taken into account that the initial phase of brain development is very different from the final one [16–18]. The reason is that in the beginning the normal children's cortex is strongly interconnected and the functioning modules are not independent, which explains the cascading effects of any early impairment on the formation and dynamics of new habits [16]. Its transformation into more and more specialized and localized as functions takes place gradually and under the influence of the constantly incoming information.

2.2 Neurophysiological organization of complex motor (praxis) functions

The ontogenesis of motor functions has a very early onset in childhood development. Like other higher functions, they depend on the dynamics of the physiological maturation of the brain in its three dimensions: vertical, horizontal and lateral, subject to the principles of heterochrony and systemicity. According to morphological studies, in the first years of postnatal ontogenesis the system of vertical connections (crust - subcortex) develops most actively, and the period of 5–6 years is a time of intensive formation of horizontal connections (intrahemispheric and interhemispheric). The levels of the projection and associative zones of the cortex also reach maturity at different time periods [19].

The formation of practical functions is determined by the stages and dynamics of motor development and its main components such as accuracy, speed and coordination. Like other higher cortical functions, they have a complex brain organization based on neural networks between a large number of sensory and motor regions of the cortex and subcortex. Despite the variety of forms, each type of praxis presupposes the execution of purposeful and consciously controlled movements with the character of automatisms. Their development at an early age is a condition for the acquisition of social habits and school skills (in particular, graphomotor). The importance of visual-motor integration and fine motor control for the formation of skills and quality of writing has been proven [5, 20, 21]. There is a large amount of evidence for the importance of visual-motor integration and fine motor control in the formation of skills and quality of writing [5, 20, 21]. Leading role in the formation of complex coordinated movements has different structures of the frontal and parietal lobes. Their maximum connectivity is the basis for the acquisition and implementation of motor habits [22].

Compared to other organs, the brain reaches adult size at a much earlier stage. Compared to other organs, the brain reaches adult size at a much earlier stage. The maturation of the cortical areas regulates the sequence and stages of development of the mental functions and abilities associated with these areas. For example, between 3 and 12 months, the increasing number of synapses in the auditory and visual cortex corresponds to the accelerated development of the child's auditory and visual perceptions. Apart from being a sensitive period for the sensory base of mental functions, the first year is associated with the active development of the motor (precentral) and kinesthetic (postcentral) areas of the cortex [23, 24].

The maturation of the leading structures for the motor functions of the frontal lobe (motor, premotor and prefrontal areas) is subject to the principle of heterochronous development. Data from neurophysiological studies show that in the first two years of life the motor areas develop most actively, and in the period 2–4 years their neuronal organization approaches that of adults. Structurally and functionally, the premotor area is close to the mature brain at 7 years of age. In the slowest maturing prefrontal cortex, several stages of significant changes in the neuronal ensemble organization are observed, which relate to the time 1 year, 3 years, 5–6 years, 9 years, and 12–14 years [25–27]. Although they do not have motor functions, the fields of the prefrontal cortex are crucial for the regulation of motor behavior. This is due to their close connection with the posterior associative cortex, the premotor cortex, the basal ganglia and the cerebellum.

Of the areas of parietal lobe, gyrus angularis and gyrus supramarginalis are those that are crucial for the development of the most complex mental functions. There is evidence for their connection with the integrative function of speech for spatially organized and visually controlled subject actions, as well as for the periods of the most significant morphofunctional changes in these regions. These periods refer to 2 and 7 years of age and coincide with a qualitative complication of the child's activities [10]. Some authors [22] consider the left supramarginal gyrus as a structure directly involved in the formation of praxis, in particular in the acquisition of motor habits, graphomotor and speech skills.

Dowell and co-authors [28] comment on the complex neurophysiological organization of praxis functions and present the structural-functional mechanism associated with the realization of learned movements. It is based on literature data, according to which the mechanism is based on the interaction of the areas of the frontal and parietal lobes. The analysis presents gyrus angularis and gyrus supramarginalis as a place for storage of spatio-temporal notions of learned movements. Due to their close connection with the structures of the premotor cortex, they have a stimulating effect on its programming functions. As a result, the premotor divisions recode the visual motor representations into motor programs and direct them to the motor cortex for execution. This leads to the conclusion that praxis functions have a universal organization, including the following main components: formation of ideas for the somatospatial and temporal characteristics of the movement (parietal cortex) and recoding of the visual image in the motor programs (premotor areas of the frontal cortex).

Because the development of praxis is based on the coding of visually perceived movement with subsequent motor imitation, a number of researchers emphasize its connection with the work of the mirror nervous system. Both forms of imitation - for known and unknown movements are related to the mechanism of comparing the currently perceived motor information with the respective motor representations. Summary data from fMRI study [29, 30] show the importance of the mirror nervous system for the early imitative behavior of children and emphasize the role of "core circuit" for imitation. It is based on the connections between three regions: the superior temporal sulcus (visual description of the action), the parietal parts of the mirror system (motor components of the action) and the frontal parts of the same (purpose of the action). Separate studies [31] have also linked gestural imitation processes to the cortical neural network of the lower frontal, anterior lower parietal and posterior upper temporal lobes, raising the idea of its bilateral organization. The latter is commented by a number of authors [32, 33], according to which, despite its bilateral organization, the imitation of the gesture has a more pronounced lateralization in the area of the left parietal cortex.

The analysis of the literature outlines the period of middle childhood (4–6 years) as sensitive to neuropsychological development. Peculiarities of motor functioning in children with typical development have important diagnostic and prognostic significance for learning readiness. However, the assessment of praxis functions during this period remains poorly developed within child neuropsychology. Systematic research in this direction faces the following tasks: outlining age trends and deriving standards for the development of praxis; development of differential diagnostic criteria for assessment of children at risk of learning disabilities; formulation of methods and approaches for preventive therapy in case of delayed formation of praxis functions. Some of these tasks we try to solve in the presented analysis of our own research.

3. Description of the research

3.1 Research objectives

The main goal of the study is to analyze the state of two types of praxis with similar brain mechanisms - dynamic and spatial postural praxis in children of preschool age (4–6 years) with typical development. The additional comparative analysis of the results aims to outline the state of the fronto-parietal neural connectivity and the developmental tendencies of the complex practical functions in the indicated age period.

3.2 Research methods

Two neuropsychological samples adapted for childhood were used to study the praxis functions - a sample for dynamic praxis and a sample for spatial postural praxis. The samples are included in the Neuropsychological Diagnostic Battery for Children [34] and are described below.

3.2.1 Test for dynamic (kinetic) praxis

It is from the group of samples for serial (successive) organization of movements. The application of the sample allows studying the following praxis components: mastering of a motor program according to a sample and automation of the program (model) with switching of the movements. Given the early age of the children, the sample includes two consecutive programs with increasing difficulty: the first alternates two elements (fist - "side"), and the second alternating elements are three (fist - "side" - palm). The movements are demonstrated by the researcher three times at a moderate pace. The instruction requires the child to memorize and repeat them six times as quickly as possible. In case of incorrect implementation, three levels of assistance are offered: first degree - re-demonstration; second degree - simultaneous performance (together with the child); third degree - simultaneous performance with verbal comment (naming the movements).

The evaluation of the performance of the two series is similar and is based on the following criteria:

- After the first demonstration - 4 points;
- After first aid (re-demonstration) - 3 points;
- After the second level of assistance (joint implementation) - 2 points;
- After the third degree of assistance (with verbal comment) - 1 point;
- Failure and after all levels of assistance - 0 points.

The analysis of the mechanisms of the dynamic praxis sample outlines its complex nature, based on the involvement of different cortical areas. Performance depends on both the development of executive functions and the acquisition of consciously controlled movements (frontal cortex) and the ability to mimic movements (lower frontal, anterior lower parietal and upper temporal lobes), deficits in which are the leading symptom in cases of developmental dyspraxia [31, 32]. Experimental data show qualitative changes in executive functions during preschool and early school age, associated with progressive growth of posterior and anterior associative fields and increased density of neural groups in the regions of the forehead [35, 36]. This defines the study of dynamic praxis as a way to assess the condition and development of the fronto-parietal nerve connections. At the same time, the heterochronous nature of neuropsychic ontogenesis hypothesizes differences in the ability to learn motor programs among typically developing children. This has great prognostic value, as it allows separating the cases of low normative performances related to the risk of learning difficulties.

3.2.2 Spatial postural praxis sample (head test)

The sample was proposed by H. Head in the early 20th century to evaluate ideomotor practice for new movements in cases of local brain damage. The variant we use was modified by Luria and defined by him as a "spatial practice of posture." Like the first, the Head test is also complex. What is specific about it is a more pronounced emphasis on the visual–spatial organization of the movements of the hand in the coordinate space of the face (horizontal, frontal, sagittal). The defining role in its implementation is played by the ideas about one's own body (body scheme) and the processes of spatial synthesis (spatial recoding), directly related to the work of the lower parietal areas. At the same time, the gestural-imitative nature of the tasks connects its neurophysiological mechanisms with bilateral fronto-parietal activity, more pronounced in the area of the left parietal cortex [28, 31].

In order to evaluate as objectively as possible, we used the sensitized version of the sample. In it, the researcher sits opposite the child and demonstrates different poses with both hands. The child should repeat them, focusing on the parts of his own body. The instruction pays special attention to the requirement that what the adult does with his right or left hand, the child must do with his right or left hand. Before the beginning of the demonstration, the child's right and left orientation on his own body and on the person sitting opposite is checked.

The demonstrated movements are divided into two groups on the principle of increasing difficulty. The first group includes 10 movements with one hand, and the second group includes 3 movements with both hands simultaneously. Bimanual movements are demonstrated only if the child completes the last three tasks of the first part (8, 9 and 10).

First group of movements:

1. The palm of the right hand on the right cheek;

2. The nape of the left hand on the left cheek;

3. The palm of the left hand on the right cheek;

4. Right hand (palm forward) rests right cheek;

5. The dorsal part of the right hand rests the chin (fingers forward);

6. The fingers of the left hand to the chin;
7. Right hand in front of the forehead (palm pointing down);
8. The palm of the left hand in front of the forehead (vertical position, facing to the right);
9. Right hand in a fist under the chin;
10. Left hand in a fist to the left cheek.

The second group of movements:

1. Left hand (palm) on the right cheek, the back of the right hand rests the left elbow;
2. The nape of the left hand is placed on the right, clenched into a fist;
3. The left hand holds the right ear; the back of the right hand is on the left cheek.

In both movements, the initial assessment is formed on the basis of the following criteria:

- Proper performance - 2 points;
- Mirror performance (spatial error type) - 1 point;
- Wrong performance (somatotopic error) - 0 points.

Despite some differences, the imitative nature of the samples for dynamic praxis and spatial postural praxis determines the common elements of their neurophysiological mechanisms. As mentioned, they are related to the formation and dynamics of complexes of the fronto-parietal nerve connections. Both samples involve preserving the spatio-temporal characteristics of visually perceived motor patterns (lower parietal divisions with more pronounced left hemispheric activity), recoding the representational images in appropriate motor programs (premotor divisions of the frontal lobe) and directing them to the motor cortex for execution. The specificity and tendencies of the performance of tasks by typically developing children in preschool age are indirect evidence of the dynamics of maturation of the cerebral mechanisms of complex praxis functions.

3.3 Participants

365 typically developing children without motor impairment signs participated in the study. All children are 4–6 years old, attend state children's schools and have Bulgarian as their mother tongue. The study considers the influence on the development of the praxis functions of three factors - age, demographic conditions (type of settlement) and gender. The following groups were formed in this connection: three age groups: 4-year-olds (116 children), 5-year-olds (128 children) and 6-year-olds (121 children); three demographic groups: - 195 children from the capital (1,500,000 inhabitance), 90 children from the big city (80,000 inhabitance) and 80 children from the small town (11,000 inhabitance). The proportion according to gender is 173 male and 192 female.

3.4 Statistics

The following statistical methods were used to process the results: three-factor analysis of variance for independent variables (F-criterion) and Post-Hoc analysis (Duncan test) to check the differences between the compared averages in the dispersion complex. The use of three-factor analysis of variance is explained by the specifics of the sample of subjects, which requires the separation of 3 independent factors - age, demographic conditions (type of settlement) and gender. For the needs of qualitative interpretation of the data, the analysis was supplemented by the percentage of types of incorrect answers when performing the tasks.

4. Results

The data from the statistical processing of the results will be presented separately for each of the samples.

4.1 Dynamic praxis test

The results of the analysis of variance showed a statistically significant influence on the state of the dynamic (serial) organization of movements and of all three factors. Statistically strongest influence was the factor age ($F = 15.62$; $p < 0.00000$), followed by the influence of the demographic factors ($F = 9.82$; $p < 0.00007$) and gender ($F = 3.89$; $p < 0.0493$). The interaction between age and demographic factors was also statistically significant ($F = 4.033$; $p < 0.003$), as was the triple interaction between age * settlement * gender ($F = 4.91$; $p < 0.00073$).

The profile of the age factor shows a regular increase in the scores of the test in the observed age period. The most significant increase is in the transition from 5 to 6 years; the differences in the average scores of children aged 4 and 5 are insignificant (**Figure 1**).

The data from the statistical check of the influence of the age factor are also confirmed by the Duncan test. It shows significant differences between the results of 6-year-olds and those of the other two age groups (**Table 1**).

The graph outlining the influence of the demographic factor shows the highest average results for children from large cities and much lower ones for children from the capital and small cities (**Figure 2**). The fact is confirmed by Duncan's test, according to which there are significant differences between the average scores in the big city and those in other settlements. There are no significant differences between the average results of the children from the capital and the small town (**Table 2**). This means that the statistically significant influence of the demographic factor is due to the very high results of the children from the big city.

The statistical influence of the gender factor is determined by the significant differences in the average results of boys and girls, where girls show better achievements in learning and performing motor programs (**Figure 3**).

The additional distribution of the results according to the different evaluation criteria outlines the trends in the development of the dynamic praxis in the period 4–6 years and supports the qualitative interpretation of the data (**Table 3**). Note that in all tables the highest values are indicated in bold.

Naturally, the weakest development of dynamic praxis is observed in children at 4 years of age. It is confirmed by the fact of the lowest performance after the first demonstration in both programs. In the two-element program, the highest results of the 4-year-olds (40%) are based on re-demonstration, and the highest in the three-element program (33%) are based on joint implementation. More than half

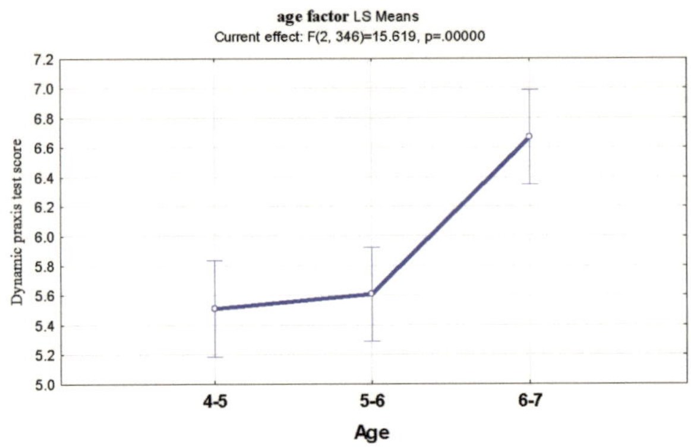

Figure 1.
Effect of age factor on the results of dynamic praxis.

Ages	{1} - 5.3043	{2} - 5.5984	{3} - 6.5574
4 years		0.163378	**0.000011**
5 years	0.163378		**0.000014**
6 years	**0.000011**	**0.000014**	

Table 1.
Significance of differences in the average scores of each age group in the dynamic praxis sample.

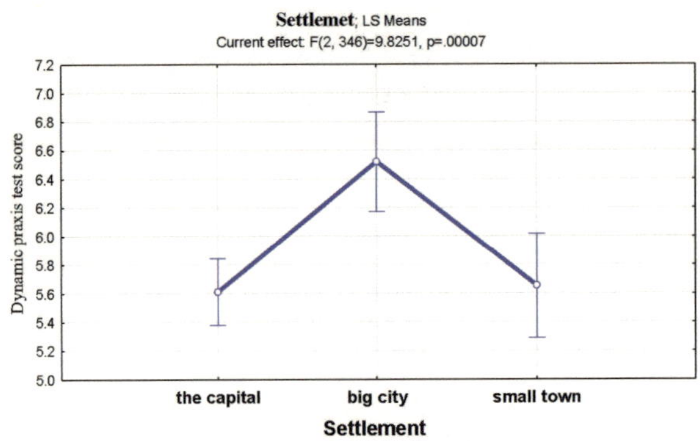

Figure 2.
Effect of demographic factor on the results of dynamic praxis.

Settlement	{1} - 5.6237	{2} - 6.4667	{3} - 5.6000
the capital		**0.000216**	0.916957
big city	**0.000216**		**0.000210**
small town	0.916957	**0.000210**	

Table 2.
Significance of the differences in the average scores between the children from the different settlements on the sample for dynamic praxis.

of the 5-year-old children (56%) master the two-element program after the first demonstration, and here too the largest number (27%) is those who master the three-element program after joint implementation. Confirmation of the positive

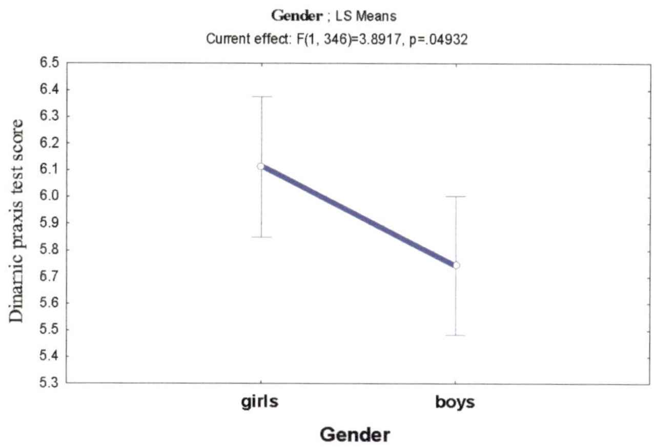

Figure 3.
Effect of gender factor on the results of dynamic praxis.

dynamics of this type of practice in both age groups gives the implementation of the program of three elements after the first demonstration: such is registered in 13% of children at 4 years and in 23% of children at 5 years. For comparison, in the group of 6-year-olds the performance of the sample after the first demonstration was respectively: 76% (in the two-element program) and 36% (in the three-element program).

4.2 Spatial postural praxis sample

In the sample for spatial postural (ideomotor) praxis, the values of the F-criterion show a statistically significant influence of the same factors: age ($F = 23.44$; $p < 0.000$), demographic conditions ($F = 8.142$; $p < 0.000$) and gender ($F = 6640$; $p < 0.010$). The double interaction age * settlement is also significant ($F = 6766$; $p < 0.000$). Similar to the previous one, in this sample the profile of the age factor has the greatest influence. It shows a gradual increase in the total score in the period 4–6 years, with a sharp rise in values in 6-year old children (**Figure 4**).

According to Duncan's test, statistically significant differences are observed between the mean scores of each of the two age groups (**Table 4**).

In both the dynamic praxis test and the ideomotor praxis test, the statistically significant influence of the demographic factor is due to the higher scores of children in the big city, followed by children in the small town and children in the capital (**Figure 5**).

This explains the existence of statistically significant differences between the results of children from the capital and the big city and children from the capital and the small town (**Table 5**). Due to the close results, there are no significant differences between the average scores of children from a big city and a small town.

Similarly to the first sample, the influence of the sex factor turned out to be, which in the sample for spatial postural praxis is again due to the higher average score of the girls (**Figure 6**).

In parallel with the cases of correct performance, in all age groups of children were analyzed the cases of mirror performance (echopraxic) and incorrect performance of the stimuli in the sample (**Table 6**). The following age trend in the distribution of the ways of performing the sample is outlined: in children at the age of 4 the cases of correct, mirror and wrong performance are distributed almost evenly (34% - 36% - 30%), with a slight predominance of the mirror performance; in children aged 5 and 6, the cases of correct implementation prevail against the

Ages	Program of 2 elements				Program of 3 elements				
	1st demonstration	2nd demonstration	joint implementation	with a verbal comment	1st demonstration	2nd demonstration	joint implementation	with a verbal comment	wrong execution
4 years	37%	**40%**	20%	3%	13%	28%	**33%**	23%	3%
5 years	**56%**	24%	12%	9%	23%	22%	**27%**	19%	9%
6 years	**76%**	16%	5%	3%	**36%**	33%	22%	7%	2%

Table 3.
Distribution of the results according to the performance criteria of the dynamic praxis test.

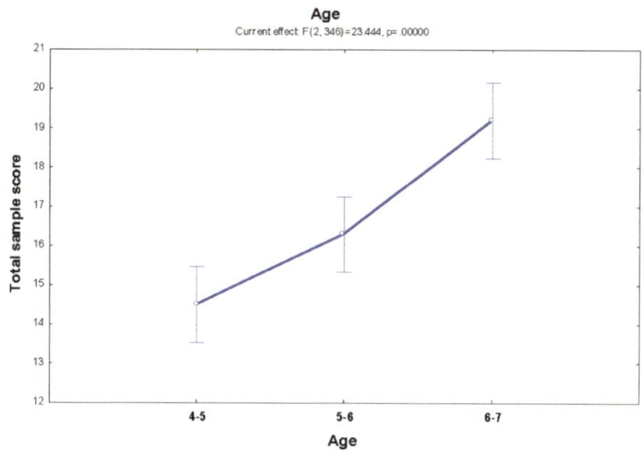

Figure 4.
Effect of age factor on the total score for spatial postural praxis ((ideomotor) praxis).

Age	{1} - 8.8696	{2} - 11.276	{3} - 16.000
4 ages		**0.011307**	**0.000011**
5 ages	**0.011307**		**0.000009**
6 ages	**0.000011**	**0.000009**	

Table 4.
Significance of differences in the average scores of each age group for spatial postural (ideomotor) praxis.

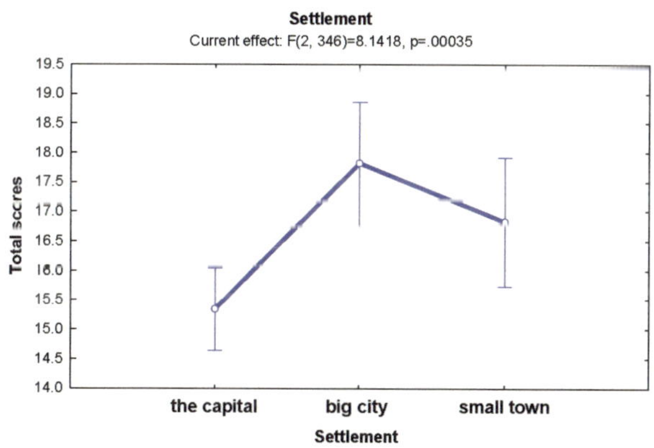

Figure 5.
Effect of the demographic (settlement) factor on the total score for spatial postural praxis (ideomotor) praxis).

Settlement	{1} - 15.402	{2} - 17.700	{3} - 16.912
The capital		**0.001127**	**0.027192**
Big city	**0.001127**		0.249477
Small town	**0.027192**	0.249477	

Table 5.
Significance of differences in the average scores for spatial postural praxis between children from different settlement.

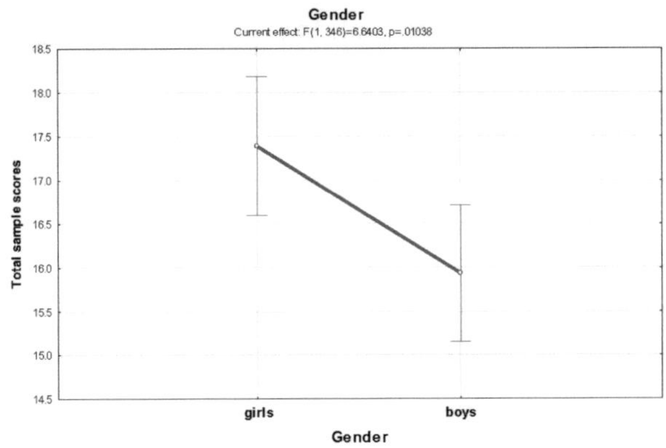

Figure 6.
Effect of the gender factor on the total score for spatial postural praxis (ideomotor) praxis).

Age	Proper performance	Mirror performance	Wrong performance
4 ages	34%	**36%**	30%
5 ages	**44%**	35%	21%
6 ages	**63%**	23%	14%

Table 6.
Distribution of the types of performance of the sample for spatial postural practice in all groups of children.

background of the reduction of the wrong, most typical for 6-year-olds, respectively: the distribution of cases of correct, mirror and wrong performance in children at 5 years is 44% - 35% - 21%, and in children at 6 years it is 63% - 23% - 14%.

5. Comparative data analysis (discussion)

The results of the study of the two types of praxis functions will be commented both sequentially and comparatively. The purpose of such a presentation is to derive the features and general trends in their development.

As mentioned, although they have their own specifics, dynamic and ideomotor practice has a similar neurophysiological organization associated with their imitative nature. Their implementation implies preservation of the spatial and temporal characteristics of the visually perceived movements (lower parts of the parietal lobe), their recoding in motor programs (premotor areas of the frontal lobe) and subsequent reproduction (motor lobes of the frontal lobe).

Statistical analysis of the data from the **dynamic praxis** sample showed a significant influence on the performance of three independent factors: age ($F = 15.62$; $p < 0.00000$), demographic conditions ($F = 9.82$; $p < 0.00007$) and gender ($F = 3.89$; $p < 0.0493$), as well as the interaction of some of them. The age factor has the strongest influence on the performance of the tasks, which is confirmed by the results of the Duncan test (**Table 1**). There were no significant differences between the groups of children aged 4 and 5, which indicates a close level of skills to perform consecutive movements. Significant differences in the results are registered between each of the indicated groups and the group of 6-year-old children, respectively: 4- and 5-year-olds ($p \leq 0.163378$), 4- and 6-year-olds ($p \leq 0.000011$) and 5- and 6-year-olds ($p \leq 0.000014$).

The data objectify the conclusion that in children with typical development the period of 4–6 years coincides with the beginning of the formation of the manual dynamic praxis. Due to the sharp improvement in praxis skills, the age of 6 years should be considered critical for the development and control of successive motor programs and the brain departments responsible for them.

Information about the dynamics in the mechanisms and stages of formation of the dynamic praxis is presented by the data from **Table 3**. They reflect the quantitative distribution of the ways of performing the tasks and the age changes in them. The conclusion that dynamic praxis is least developed in 4-year-old children is confirmed, most of whom (39%) perform the two-component program after the second demonstration, and one third (33%) master the program of three movements only in conditions of joint implementation. The positive changes in 5-year-olds are mainly related to the improved implementation of the two-element program, as more than half of the children (56%) implement this program after the first demonstration. At the same time, the three-element program continues to be dominated by the joint implementation criterion (27%). Despite the fact that at the age of 5 the number of children who mastered the three-element program after the first demonstration increased, the results of the criterion for joint implementation still prevailed. Significant changes in the state of dynamic praxis are registered after the age of 6, which is confirmed by the growing number of performances after the first demonstration: 76% implementation of the two-element program and 36% implementation of the three-element program. There is also a significant reduction in cases of joint implementation (5%) and especially the performance with verbal comment (3%).

The presented data provide indirect information about the stage of formation of the functional system of dynamic praxis. They show that at the age of 4 years the brain structures associated with the realization of motor series are organized on a generalized principle, which after the age of 5 begins to be replaced by a process of gradual specialization. The significant change in the results during the period 6–7 years is a reason to consider this age as sensitive for the development of the dynamic praxis and the formation of the bilateral fronto-parietal neural complexes.

According to the influence of the demographic factor, the best development of the dynamic praxis is shown by the children from a big city, significantly ahead of those from the capital and the small town (**Table 2**). To some extent, this did not confirm the expectation of a leading place for children from the capital in terms of neuropsychological development. Although the facts need further study, it can be assumed that in contrast to the moderately populated, places with a very high concentration of population do not have the necessary stimulating effect on the cerebral ontogenesis of motor and executive functions of children. In our opinion, the causes are complex, including a variety of factors with different effects on early cerebral ontogenesis. This corresponds to the cited theories of the specific interaction of biological and social factors and the impact of the environment on the exposure to genetically determined heterochrony, leading to a variety of individual variants of development.

Although less pronounced, the statistically significant influence of the gender factor is due to the higher results of girls, and their presence indicates the connection of this factor with the development of complex praxis functions. This conclusion is complemented by the similar influence of gender on the performance of the sample for spatial postural praxis. Therefore, the fronto-parietal nerve connections and the mirror nervous system of girls undergo faster development, the effect of which may have preferences for various manual activities.

Qualitative analysis of the results leads to the following conclusions: available for children at 4 years is the shortened version of the sample for dynamic practice, while the implementation of the three-element version is associated with many gaps and motor perseverations; 5-year-olds do better with the complex version of the test, but the transcoding of spatio-temporal representations in motor programs is slow, movements are stiff and require maximum concentration; in children at the age of 6 the recoding of motor representations is significantly improved, the performance becomes more accurate, there is an opportunity for self-control and correction of errors.

Statistical analysis of the **ideomotor praxis** sample showed a significant influence of the same three independent factors: age ($F = 23.44$; $p < 0.000$), demographic conditions ($F = 8.142$; $p < 0.000$) and gender ($F = 6640$; $p < 0.010$). The leading influence of the age factor is again related to the gradual improvement of the results of the tasks and to the presence of statistically significant differences between each of the two age averages (**Table 4**). This is complemented by the quantitative distribution of data on the individual criteria for sample performance: Proper performance, Mirror performance and Wrong performance (**Table 6**).

The observed age trend is associated with a transition from a predominant mirror performance in children at 4 years (34%) to proper performance in children on 5 and 6 years. The close percentage results of the three types of performance at the age of 4 years speak of a generalized principle of organization of the brain mechanisms, characteristic in the performance of the test for dynamic praxis. It is replaced by processes of gradual specialization of the motor areas related to motor imitations (parietal and frontal) and leads to an increased number of proper performance in the next two age periods (44% in children at 5 years and 63% in children at 6 years). It can be assumed that the reduction of the cases of mirror and wrong execution is directly related to the active maturation of the lower parietal departments as responsible for the spatial synthesis and the ideas about one's own body. The age dynamics in the skills of children to imitate movements gives grounds to consider the age of 6 years as a sensitive period for the development of spatial postural (ideomotor) praxis.

Valuable information about age-related changes in visual-spatial orientation and ideomotor practice is provided by the comparison of each of the two age groups according to the criteria for correct, mirror and wrong performance, conducted by Student's t-test. The data show that according to the criterion for correct performance significant differences are registered between each of the two age groups: 4- and 5-year-olds ($p \leq 0.001$); 4- and 6-year-olds ($p \leq 0.001$) and 5- and 6-year-olds ($p \leq 0.001$), and their presence indicates a uniform and gradual formation of the mechanisms of spatial postural praxis in the considered age period.

According to the criterion for mirror (echopraxical) performance, the picture of the results is different due to the lack of significant differences between the groups of 4- and 5-year-olds ($p \geq 0.05$). There are significant differences in the cases of mirror performance of tasks between children aged 4 and 6 ($p \leq 0.001$), as well as between children aged 5 and 6 ($p \leq 0.001$). This means that only after the age of 6 do most of the children become able to perform mental spatial recoding of the motor image and adequate spatial synthesis of the observed movements. As the mirror performance is explained by underdevelopment of the spatial orientation, the close values according to this criterion in the first two age groups (4 and 5 years) confirm the sensitive nature of the 6-year-old age and for the development of the spatial function.

Similar to the first criterion, significant differences in the criterion for incorrect (wrong) performance of the sample are registered between each of the two age groups, respectively: between children aged 4 and 5 ($p \leq 0.001$), between children aged 4 and 6 ($p \leq 0.001$) and between children aged 5 and 6 ($p \leq 0.001$). Although it decreases with age, the presence of these cases indicates an incomplete process of formation of the scheme of the body (somatognosis), directly related to the spatial orientation in its parts (on oneself and on others). The presence of such somatotopic errors is more global in nature and is a serious indicator of future learning difficulties.

Qualitative changes in the analyzed executive functions in the period 4–6 years are explained in some neuroanatomical data showing increased growth of the posterior and anterior associative fields and increased density of neural groups (ensembles) in areas of the frontal lobe, in particular in the premotor cortex [35, 36]. Developmental changes in the bioelectrical activity of the child's brain, related to the predominant alpha rhythm and improvement of its spatial organization, are also indicated as a sign of maturation of the cerebral departments [37]. The age periods 6–7 and 9–10 years are indicated as transient for the dynamics of the alpha rhythm, which supports the conclusion about the importance of 6 years of age in the development of praxis functions.

The demographic factor has a significant impact on ideomotor praxis, which again is due to the highest average score for children from a large city and the lowest for children from the capital. Statistically significant on the Duncan test are the differences between the averages of the children from the capital and from a big city ($p \leq 0.001127$), as well as children from a small town and a big city ($p \leq 0.027192$). The similar results for the influence of the demographic factor on the two types of praxis functions in the considered period confirm the need for more in-depth research on the relationship of social factors and neuropsychological development in childhood.

The results for the influence of gender on the spatial orientation and ideomotor praxis are similar to those in the dynamic praxis sample and confirm the conclusion for faster maturation of the mirror nervous system and fronto-parietal neural ensembles in girls.

The specificity of the samples confirms the action of the heterochronous principle of neuropsychic development. The main evidence for this is the different dynamics of the formation of the studied functions, related to the faster development of the neurophysiological organization of the spatial postural praxis in comparison with that of the dynamic praxis. One of the reasons is the slower maturation of the structures of the frontal lobe, responsible for recoding and realization of the spatio-temporal parameters of the complex serial movements. This is confirmed by the results for correct performance of the two groups of tasks in in children at 6 years: 63% for ideomotor praxis and 36% for the complicated variant of dynamic praxis.

The observed age trend shows the variety of individual differences in children with typical development, as well as the fact that a large part of them enter school with insufficiently developed praxis skills.

6. Conclusion

The state of praxis functions in preschool has important diagnostic and prognostic significance for child development. Their implementation involves preserving the spatial parameters of visually perceived motor models with subsequent recoding in motor programs, which makes it an objective criterion for the formation and dynamics of fronto-parietal neural networks and structures of the mirror nervous system. The formation of complex praxis functions is influenced by three independent factors - age, demographic conditions and gender. The leading role of the age factor proves the determining effect of neurobiological changes on the neuropsychological development of the child. The leading role of the age factor proves the determining effect of the dynamics of neurophysiological changes on the neuropsychological development of the child. The heterochronous principle to which this development is subject explains the uneven nature of the formation of praxis functions, in particular those of dynamic and spatial postural praxis. Another reflection of it is the great variability and diversity in the rate of maturation of the brain departments responsible for the realization of these functions.

The influence of the age factor is related to qualitative changes in the motor skills of children, most pronounced at the age of 6 years, which defines it as critical for the formation of complex praxis functions. The variability and individual dynamics of neuropsychological development determine the differences in the functioning of the practice and objectify the need for its inclusion in the complex assessment of children. The registration of cases of delayed formation of praxis functions in the preschool period will lead to the development of stimulant therapy and overcoming future learning difficulties.

Conflict of interest

The authors declare that there is no conflict of interest.

Author details

Neli Cvetanova Vasileva[1*] and Jivko Dimitrov Jekov[2]

1 Sofia University "St. Kliment Ohridski", Sofia, Bulgaria

2 Medical University "Prof. Dr. Paraskev Stoyanov", Varna, Bulgaria

*Address all correspondence to: vasnel@abv.bg

References

[1] Glozman ZhM. Nejropsixologiya detskogo vozrasta. Moskva: Akademiya; 2009

[2] Xomskaya ED. Lateral'naya organizaciya mozga kak nejropsixologicheskaya osnova tipologii normy`. V: I Mezhdunarodnaya konferenciya pamyati A. R. Luriya (Xomskaya, Axutina, red). Moskva.1998. s:138-144

[3] Axutina TV. Nejropsixologicheskij podxod k diagnostike i korrekcii trudnostej obucheniya pis'mu. V: Sovremenny'e podxody` k diagnostike i korrekcii rechevy'x rasstrojstv. Sankt-Peterburg: SPb Universitet; 2001. s:195-213

[4] Tupper DE, Cicerone K. Introduction: Developmental and Rehabilitative issues in the Neuropsychology of everyday life. The Neuropsychology of everyday life (Tupper, Cicerone, eds.). Kluwer Academic Pub Group. 1991. pp. 3-14

[5] Piek J, Hands BP, Licari M. Assessment of motor functioning in the preschool period. Neuropsychol Rev. 2012; **22 (4)**:402-413. DOI: org/10.1007/s11065-012-9211-4

[6] Colombo-Dougovito AM. The role of dynamic systems theory in motor development research: how does theory inform practice and what are the potential implications for autism spectrum disorder? International Journal on Disability and Human Development. 2017; **16(2)**:141-155

[7] Gabbard C, Caçola P, Bobbio T. Studying motor development: A biological and environmental perspective: In E. Kahraman and A. Baig (Eds.), Environmentalism: Environmental strategies, and environmental sustainability. Nova Science Publishers. 2009. pp. 129-139

[8] Hadders-Algra M. The Neuronal Group Selection Theory: a framework to explain variation in normal motor development. Developmental Medicine & Child Neurology. 2000; **42**:566-72

[9] Mountcastle VB. The columnar organization of the neocortex. Brain. 1997; **120(4)**:701-722

[10] Mikadze YuV. Nejropsixologiya detskogo vozrasta.Sankt-Peterburg: Piter. 2014

[11] Axutina TV, Py'laeva NM. Nejropsixologicheskij podxod k korrekcii trudnostej obucheniya. V: Nejropsixologiya segodnya. Moskva: MGU. 1996. pp:160-170

[12] Manelis NG. Nejropsixologicheskie zakonomernosti normal'nogo razvitiya. Shkola zdorov'ya. 1999; **6 (1)**:8-24

[13] Mikadze YuV. Diferencial'naya nejropsixologiya detskogo vozrasta. Voprosy` psixologii. 2002, **4**:111-119

[14] Hartlage LC. Introduction to the Neuropsychology of individual differences. In: The neuropsychology of individual differences (Hartlage & **Telzrow,** eds.). Springer-Science. 1985. pp. 1-22

[15] Karmiloff-Smith A. Development itself is the key to understanding developmental disorders. Trends in Cognitive Science. 1998; **2**:389-398

[16] Karmiloff-Smith A. Crucial Differences between Developmental Cognitive Neuroscience and Adult Neuropsychology. Developmental Neuropsychology. 1997; **13 (4)**:513-524

[17] Huttenlocher PR, Dabholkar AS. Regional differences in synaptogenesis in human cerebral cortex. Journal of Comparative Neurology. 1997; **387:**167-178

[18] Johnson MH. Functional brain development in humans. Nature Reviews Neuroscience. 2001, **2**:475-483

[19] Farber DA, Dubrovinskaya NV. Mozgovaya organizaciya kognitivny'x processov v doshkol'nom vozraste. Fiziologiya cheloveka. 1997; **23 (2)**:25-32

[20] Volman MJ, van Schendel BM, Jongmans MJ. Handwriting difficulties in primary school children: a search for underlying mechanisms. American Journal of Occupational Therapy 2006, **60**:451-460

[21] Van Hartingsveldt MJ, De Groot IJM, Aarts PBM, Nijhuis-van der Sanden MWG. Standardized tests of handwriting readiness: a systematic review of the literature. Developmental Medicine and Child Neurology 2011, **53(6)**:506-515. DOI:10.1111/j.1469-8749.2010.03895.x

[22] Ramachandran VS. Mozg rasskazy'vaet. Chto delaet nas lyud'mi. Moskva: Kar'era Press. 2015

[23] Semenova LK, Vasil'eva VV, Cexmitrenko TA. Strukturny'e preobrazovaniya kory` bol'shogo mozga cheloveka v postnatal'nom ontogeneze. V: Strukturno-funkcionalnaya organizaciya razvivayushhegosya mozga. Leningrad: Nauka. 1990. pp. 8-44

[24] Berk LE. Child development (6 ed.). Pearson Education Inc. 2003

[25] Goldberg È. Upravlyayushhij mozg. Lobny'e doli, liderstvo i civilizaciya. Moskva: Smy'sl; 2003

[26] Shumejko NS. Razvitie sensomotornoj kory` bol'shogo mozga cheloveka i stanovlenie dvigatel'ny'x funkcij v ontogeneze. Al'manax„Novy'e issledovaniya". 2004; **1-2**:432-443

[27] Maryutina TM. Vozrastnaya psixofiziologiya. V: Psixologiya razvitiya (Marcinkovskaya red.). Moskva: Akademiya. 2005. s: 91-188

[28] Dowell LR, Mahon EM, Mostofsky SH. Associations of Postural Knowledge and Basic Motor Skill with Dyspraxia in Autism: Implication for Abnormalities in Distributed Connectivity and Motor Learning. Neuropsychology. 2009; **23(5)**:563-570. DOI: 10.1037/a0015640

[29] Iacoboni M, Dapretto M. The mirror neuron system and the consequences of its dysfunction. Nature Rev. Neurosci 2007; **7**:942-951

[30] Rizzolatti G, Fabbri-Destro M. Mirror neurons: From discovery to Autism. *Experimental Brain Research 2010; 200* **(3-4)**:223-237. DOI:org/10.1007/s00221-009-2002-3

[31] Aziz-Zadeh L, Koski L, Zaidel E, Mazziotta J, Iacoboni M. Lateralization of the human mirror neuron system. J Neurosci. 2006; **26(11)**:2964-70. DOI:10.1523/JNEUROSCI.2921-05.2006

[32] Mühlau M, Hermsdörfer J, Goldenberg G, *et al.* Left inferior parietal dominance in gesture imitation:an fMRI study. Neuropsychologia. 2005; **43(7)**:1086-98DOI:10.1016/j. neuropsychologia.2004.10.004

[33] Molenberghs P, Cunnington R, Mattingley JB. Is the mirror neuron system involved in imitation? A short review and meta-analysis. Neurosci BioвehavRev. 2009; **33(7)**: 975-980. DOI:10.1016/j.neubiorev.2009.03.010

[34] Axutina TV, Polonskaya NN, Py'laeva NM, Maksimenko MYu. Metodiki nejropsixologicheskogo issledovaniya detej. Nejropsixologicheskaya diagnostika, obsledovanie pis'ma i chteniya mladshix shkol'nikov (pod red. Axutinoj, Inshakovoj). Moskva: Sekachev. 2012. s: 11-33

[35] Semenova OA, Koshel'kov DA, Machinskaya RI. Vozrastny'e

izmeneniya proizvol'noj regulyacii deyatel'nosti v starshem doshkol'nom i mladshem shkol'nom vozraste. Kul'turno-istoricheskaya psixologiya. 2007; **4**:39-49

[36] Kirkwood MW, Weiler MD, Holmes-Bernstein J, *et al.* Sources of Poor Performance on the Rey-Osterrieth Complex Figure Test among Children with Learning Difficulties: A Dynamic Assessment Approach. The Clinical Neuropsychologist. 2001; **15(3)**:345-56

[37] Alferova VV, Farber DA. Otrazhenie vozrastnykh osobennostej funkcional'noj organizacii mozga v ehlektroehncefalogramme pokoya. V: Strukturno-funkcional'naya organizaciya razvivayushchegosya mozga. Leningrad: Nauka. 1990. s. 45-64

Section 2

Pathophysiology and Neuroplasticity

Chapter 3

Cerebral Damage after Stroke: The Role of Neuroplasticity as Key for Recovery

Mubarak Muhammad and Tasneem Muhammad Hassan

Abstract

Stroke remains global health care problem that constitutes world's second-leading perpetrator of mortality and third most pronounced cause of all disabilities. The hallmark of cerebral stroke is the persistent loss of cerebral function consequence of abnormality of the blood supply. The ultimate goal of stroke care is to recover and maximize the cerebral functions lost due to the cerebral damage. Therefore, understanding the mechanism of cerebral damage after stroke is fundamental to comprehension of mechanisms of recovery following stroke, as well as key towards eliminating devastating human disability as a result of stroke. Therapeutic strategies aim to harness and enhance neuroplasticity offers reasonable level of hope towards maximizing recovery from post stroke impairments. This paper therefore, highlighted the mechanism of cerebral damage after stroke as well as elucidates the concept of neuroplasticity as key for recovery following stroke.

Keywords: cerebral cortex, cerebral damage, stroke, neuroplasticity, stroke recovery

1. Introduction

Stroke also known as cerebrovascular accidents is the world's second death-perpetrating disease after cardiovascular diseases [1, 2], and it affects about 13.7 million people annually in the globe [3]. About one third of all strokes translate into fatalities, and another one third constitutes stroke survivors staying with residual disability that accounts as foremost noticeable root of long-term neurological disability in adults [4, 5] and third most common cause of all disabilities globally [6]. Stroke classically depicts a syndrome with sudden onset of acute focal injury of the central nervous system (CNS) of vascular origin that produces focal or global neurological deficit in accordance with affected area of blood supply [7]. Thus, based on the isolated territory of the brain involve, stroke can be cerebral stroke, brainstem stroke, cerebellar stroke, or thalamic stroke, while based on underline cause it can be ischemic stroke (thrombotic, embolic, lacunar, watershed, or cryptogenic) which results from brain vascular occlusion, or hemorrhagic stroke (intraparenchymal or subarachnoid) which is due to blood-related aberrations [8].

Cerebral stroke results in loss of cerebral cortex related functions that manifests as motor impairment [9–11], sensory impairment [12–14], cognitive impairment [15–17], balance impairment [18] among others. The motor function of the cerebral cortex is embedded in the motor cortex (primary motor area, premotor cortex, supplementary motor area, cingulate motor areas) located in the frontal lobe anterior to central sulcus, the motor cortex is responsible for planning, initiation, execution, and regulation of voluntary movement which is achieved through originating descending corticospinal tract and corticobulbar system to the spinal cord and brainstem respectively [19]. Cerebral cortex plays principal role in sensory/perceptual functions by providing meaning to all sensations (except sense of smell) through primary somatosensory cortex in the postcentral gyrus of the parietal lobe, and other primary cortical sensory areas such as auditory cortex in the temporal lobe and visual cortex in the occipital lobe. Cognitive function involves multifaceted domains of cognitive processes including memory, learning, attention, thought, comprehension, perception, language among others [20]. Each of these domains of cognition requires cerebral cortex, illustration can be seen in memory domain where memory acquisition involves sensory cortex, memory retrieval involves prefrontal cortex, and memory storage is distributed throughout the cortex [21]. Balance and coordination of movement involve integrated functioning of both pyramidal and extra-pyramidal systems, and the cerebral cortex is the main principal origin of pyramidal system.

The mechanism of cerebral damage after stroke determines the cerebral stroke impairments, and the mechanism of damage is relative to whether the type of stroke is ischemic or hemorrhagic. Ischemic stroke consists of five distinct pathophysiologic mechanism each of which has distinct time frame; these includes immediate (within minutes) peri-infarct depolarization and excitotoxicity, hours later by neuro-inflammation and oxidative stress, days later by apoptosis [8]. In addition to ischemia related cascade of events aforementioned, hemorrhagic stroke is associated with two additional unique pathophysiologic phases. The primary; acute phase which is due to physical effect of hematoma (mass effect) from the mass accumulated blood, and the secondary; subacute phase termed as cytotoxicity from secondary metabolites of blood components [22–24].

Recovery to some extent from post stroke impairments observed among stroke survivors was one of the early evidences that led to move away from outdated dogma widely misconceived previously that; there was no possibility for repair or change within the CNS after it had suffered a lesion; and that once there is damage such as stroke that leads to neuronal demise inadvertently, the brain structures and functions are lost forever [25, 26]. It is now well-established fact that CNS repair or change itself but it just that it relatively does not do well enough, and that functional recovery after damage relies on neuroplasticity [27, 28]. Neuroplasticity is life-long natural capability of the CNS to rearrange itself in both molecular form and function in response to new experience or stimulus. Brain plasticity is pivotal to functional recovery after cerebral stroke, and this spontaneous, endogenous and intrinsic capacity of the brain is what restorative rehabilitation approaches for stroke explore, promote and remodel in the right direction to achieve optimal functional recovery after stroke [29, 30].

There is exploding surge among scientists to pay more attention in searching for various therapeutic strategies that can enhance neuroplasticity to augment functional recovery with rehabilitation after stroke [31–34]. Although this strategy is still in developmental stage but the reasons for this shift in attention are not far-fetched. Firstly, the thrombolytic/thrombectomy clinical treatment available for acute stroke has a very restrictive time window of administration of 4–5 hours of lesion onset [35]. This is in contrast to restorative/rehabilitative interventions that has unlimited therapeutic window of lifelong applicability [36]. Secondly, rehabilitation interventions are still far from sufficiency for optimal and ideal recovery from impairments after stroke [37], as about 50% of stroke survivors still leaves with residual disability and remain functionally dependent despite rehabilitative management [38]. Understanding the mechanisms of cerebral damage and their recovery after cerebral stroke is essential towards development of strategies that harness and enhance neuroplasticity in combination with rehabilitation processes [39]. This paper therefore discusses the mechanism of cerebral damage after stroke as well as elucidates the concept of neuroplasticity as key for recovery following stroke.

2. Mechanism of cerebral damage after stroke

In ischemic stroke, irreversible cascade of damage to the brain tissue ensue once the cerebral blood flow (CBF) reduces to less than 12 ml/100 g/min of the normal range of 50–60 ml/100 g/min. Within seconds of this abrupt ischemic insult, neuronal cells in the center of ischemic region termed as ischemic prenumbra undergoes anoxic depolarization due to loss of ATP-dependent ionic pump homeostasis, and they never repolarize [40]. This necrotic core of ischemic prenumbra is enclosed by a zone of relatively lesser impacted tissue termed as ischemic penumbra, which is abridged functionally silent by the reduced blood flow but maintains metabolically active and therefore can repolarize at the expense of further energy consumption [41]. This repetitive depolarization and repolarization of ischemic penumbra are termed peri-infarct depolarization and the important period of time during which this volume of brain tissue is salvageable is referred to as the window of opportunity. The energy failure in the functioning of ATP dependent sodium potassium pump in the ischemic prenumbra results in massive uncontrolled anoxic depolarization that results in opening of voltage-gated calcium channels, mitochondrial dysfunction which further deplete energy required to maintain ion gradient, and abnormally extracellular buildup of excitatory amino acids [42, 43].

Consequently, excitatory glutamate and other excitatory amino acids such as aspartate becomes excessively released, and glutamate hyperexcitation of glutamate N-methyl-D-aspartate (NMDA) receptor, which is arguably the most calcium-influx allowing ionotropic glutamate receptor; results in massive influx of calcium ion (Ca^{++}) into hypoxic neuron. Calcium ion triggers series of cascading events that ultimately lead to neuronal demise through activation of proteolytic enzymes, stimulation of pathogenic genes, lipid peroxidation and free radical generation [44]. For this; glutamate and other excitatory amino acids are cumulatively termed excitotoxins, and their accompanying neuronal damage termed excitotoxicity [45]. Calcium activates key number of disparaging intracellular enzymes such as proteases, kinases, lipases, and endonuclease that not only wildly permits release of cytokines and other mediators that result in the loss of cellular integrity but also orchestrated triggering of intrinsic apoptotic pathway of neuronal death. Specifically, calcium through mobilizing phospholipases hydrolyses membrane bound glycerophospholipids to yield free fatty acids, which enable free radical peroxidation of other membrane bound lipids. Calcium through mobilizing proteases lyses integral structural proteins and activates nitric oxide synthase enzyme that triggers free radical machinery [46].

Prior excitotoxicity activates microglia and astrocytes which are the brain resident innate immunity to reacts and release cytokines, chemokines (chemotaxis cytokines), and matrix metalloproteases (MMPs). This constitutes neuro-inflammation, and microglia activation institutes the initial vital neuro-inflammatory response in acute stroke, which together with blood-borne innate immune cells and later adaptive immune cells support the course. This neuro-inflammatory response supposedly aims to reduce injury processes but this response under stroke pathology develops improperly more reactive and aggressive to yield numerous inflammatory mediators that trigger apoptosis and orchestrate lethal neuronal injury [47, 48]. Activated microglia becomes phagocytes that can release plethora of substances, some of which are neuroprotective such as neurotropic factors; nerve growth factor (NGF), brain-derived neurotrophic factor (BDNF), insulin-like growth factor I (IGF-I), and growth associated protein (GAP-43/B-50), while some are neurotoxic such as tumor necrosis alpha (TNF-α), interleukin-1β (IL-1β), and interleukin-6 (IL-6). Blood–brain barrier (BBB) which confers brain with protection against systemic toxins is disrupted by matrix metalloproteinases (MMPs) with MMP-2 (gelatinase A) and MMP-9 (gelatinase B) being the leading concerns in cerebral ischemia [49]. MMP-2 that is normally expressed at low levels becomes increased during cerebral ischemia to galvanizes MMP-9, which abolishes components of the basement membrane in the vascular wall leading to BBB distraction, thus allowing further infiltration of inflammatory mediators and other potential toxins [50].

Oxidative stress signifies disparity in the high-level oxidants (free radicals) with respect to corresponding nonconforming low level of antioxidants. Long term cerebral hypo-perfusion produces abnormal proportions of reactive oxygen species (ROS) and/or reactive nitrogen species (RNS) oxidants through several mechanisms of injury, such as mitochondrial inhibition, calcium ions overload, ischemia–reperfusion injury, and neuroinflammation [51]. During cerebral ischemia, there is mitochondrial inhibition of oxidative phosphorylation due to the lack of sufficient oxygen, and the oxygen depleted cell shift to glycolytic pathway of ATP generation that results in lactate and hydrogen ion (H^+) build-up in the mitochondria and the consequent reversal of the H^+ uniporter on the mitochondrial membrane that results in superfluous cytosolic H^+ buildup and acidosis [52]. Acidosis partly lead to oxidative stress by supplying excessive H^+ for the successive progression in the generation of hydrogen peroxide (H_2O_2) and the final hydroxyl radicals (•OH) either in the turnout of transition metal ions (Fenton reaction) or in the presence of superoxide radical (Haber-Weiss reaction), with this effect more pronounced in neurons due to inherently low anti-oxidant defense. In addition, the compelling protein and lipid oxidant peroxynitrite ($OONO^-$) of RNS is favorably generated in the oxygen depleted cell by the reaction of nitric oxide (NO) and superoxide (O_2•−), thereby also contributing to oxidative stress.

Calcium overloads, as a result of glutamate mediated NMDA receptor excitotoxicity, also contributes in neuronal oxidative stress at cytosolic and mitochondrial level. At cytosolic level, excessive calcium ion activation of key intracellular enzymes such as neuronal nitric oxide synthase (nNOS) via Ca^{2+} binds calmodulin to induce subsequent downstream effect, as nNOS catalysis results in generation of nitric oxide (NO) free radical from L-arginine [53, 54]. At the mitochondrial level, excessive calcium ion influx into mitochondrial matrix leads to the inner mitochondrial accumulation of momentous level of Ca^{2+} via mitochondrial calcium uniporter (MCU) which proliferates disturbance of usual bio-energetic, mitochondrial ROS, and membrane permeability [55].

Apoptosis is a physiological mechanism of cell death through programmed cellular machinery of either extrinsic or intrinsic pathways [56]. Under stroke pathology, neuronal demise by necrosis preponderance in the ischemic prenumbra is marked by excitotoxicity, while additional process of neuronal demise by apoptosis which is more delayed and predominant in the ischemic penumbra occur in a fashion where apoptosis becomes dysregulated [57]. Thus, while the neurons within the core infarct die by immediate necrosis due to insufficient ATP, the penumbra die by ATP requiring process of apoptosis, supporting the established evidence that cellular demise after cerebral ischemia transpires through both necrosis and apoptosis [58]. Multiple pre-existing pathophysiologic mechanisms that can induce apoptosis after cerebral ischemia includes pro- calcium influx, pro-inflammatory cytokines and oxidative stress [59]. Apoptosis can be caspase-dependent or caspase-independent, and the most common is caspase-dependent which is initiated and triggered through distinctively intrinsic (or mitochondrial) pathway or extrinsic (or death receptor) pathway. Both intrinsic and extrinsic pathways share similar terminal phase termed execution phase where caspase 3 leads to the destruction of cellular components and cell death [60].

In hemorrhagic stroke, the mechanism of damage begins with additional process of mass effect from the mass accumulated blood, and cytotoxicity from the secondary metabolites of blood components, in addition to shared common damaging caused by ischemia such as excitotoxicity, neuroinflammation, oxidative/nitrosative stress, and apoptosis. The initial bleed from the cerebral hemorrhage causes immediate physical disruption of the cellular cytoarchitecture of the brain and increases local pressure which can cause compressions, hypothetically disrupting blood flow and principally causing brain herniation [61]. The subsequent expansion of hematoma causes mass effect of hematoma growth leading to further rise in intracranial pressure, brain herniation, and impacted blood flow that is correlated with neurologic deterioration and degraded clinical outcomes. Depending on the dynamic of hematoma expansion (growth), the primary damage ensues within minutes to hours subsequent to the onset of bleeding and is basically due to mechanical damage associated with the mass effect [62].

Secondary injury after cerebral hemorrhage termed as cytotoxicity occurs due to series of events initiated by the prior primary injury mechanism (mass effect), that is specifically due to body response to the hematoma for instance inflammatory response, and from the multiple blood components released from hematoma [61]. The extravasated blood components released from hematoma being implicated to cumulatively imposed cellular toxicity includes; majorly the erythrocytes and plasma proteins, and the damage-associated molecular patterns (DAMPs) which are nucleic acids, extracellular matrix components, proteins, lipid mediators, ATP and uric acid released from necrotic tissues [63]. At the early stage of cytotoxicity, the toxicity of extravasated blood plasma components such as coagulation factors, complement components, and immunoglobulins are known to be the main contributing factor of cellular damage. Subsequently, erythrocytes lysis leads to release of its major intracellular component hemoglobin (Hb), which when metabolize via hemoglobin metabolic pathway release degradation products; heme and iron (Fe). Both Hb and its degradation products are potent cytotoxic chemicals capable of causing death to many brain cells through mechanism of free radical generation with substantial increase oxidative stress and subsequent damage to DNA [62].

3. Concept of recovery after post stroke cerebral damage

The ultimate goal of stroke management is to promote optimal recovery of lost functions and reduce further injury. This recovery depends majorly on brain plasticity; a spontaneous regeneration process that encompasses neural plastic changes in the lesioned hemisphere to reestablish its structural and functional reorganization. Brain plasticity under pathological condition completely differs from plasticity under properly functioning brain. For instance, plasticity in normally functioning brain is a prerequisite basis of learning and memory that involves plastic adaptation such as long-term potentiation (LTP). This is opposed to plastic changes observed using MRI in cerebral stroke pathology, that involves modification in intracortical myelin, augmented neurogenesis, improved spine density in neuronal dendrites and alterations in astrocyte volume [64].

Stroke recovery to certain extent also depends on severity extent of the initial injury deficit as the severity of the damage is inversely related to the prognosis for recovery [65]. But it was also observed that recovery differs even among post stroke patients with similar clinically assessed severity. This apparently stress the recovery role of other brain endogenous survival mechanism such as extent to which collateral circulation bypass to supply blood to the perilesional neurons, angiogenesis, inhibitory neurotransmitters that counteract excitotoxicity, and multiple representations of the same function in different cortical areas [66]. Appropriate rehabilitation and drug treatment that target underline cause of stroke are also critical to recovery after post stroke cerebral damage. Rehabilitation aims to maximize optimum recovery of lost functions as a result of impairments deficit after stroke but overall, brain plasticity underlies recovery promoted by rehabilitation [67–69].

Recovery from stroke has also been attributed to be dependent on resolution of early local processes in the brain that includes resolve of perilesional edema, re-emergence of circulation within the ischemic penumbra, resolution of remote functional depression of neurological function induced by process of diaschisis [70]. As previously stated stroke recovery majorly depends on brain reorganization process of plasticity which in turn dictates recovery promoted by rehabilitation. Mechanism through which rehabilitation mediates brain plasticity to promote recovery has been studied and explained. Rehabilitation such as physical therapists stroke interventions modifies neurotrophic factor expression in the CNS especially brain derived neurotrophic factor (BDNF), which in turn upon binding with its tyrosine kinase B (TrkB) cognate receptor recruits a cascade of signaling pathways that ultimately mediates activity-associated plasticity of neurons [71, 72]. Activity-associated plasticity signifies a means of functional and structural neuroplasticity that is tailored by the depolarizing behavior of neurons, and the mechanisms governing activity-associated plasticity includes LTP and activity-associated development of corticospinal circuitry among others [72]. Therefore, through brain plasticity after cerebral stroke, reorganization by recruiting cortical or subcortical structures to adopt the function of the injured tissue, reinforcement of remaining synaptic pathways and then creating new connections, recruitment of other pathways that are functionally alike the damaged tissue but anatomically distinct, strengthening of existing but weaker and functionally silent connections, can all be achieved to recover lost cerebral functions [73].

4. Neuroplasticity and its basic physiology

Neuroplasticity is a general term that covers all available processes of neuronal reorganization possible [66], such as neurogenesis, synaptogenesis, dendritic arborization, axonal sprouting, LTP, recruitment of other pathways, reinforcement of functionally silent synapses. Neurogenesis is the process of generating of neurons of neural cell types from precursors neural stem cells and/or neural progenitor cells (NPCs) [74]. Synaptogenesis is a broad term that encompasses the complex process of synaptic contacts formation, maturation and maintenance which form the basis for establishing neural circuits [75]. Dendritic arborization describes a process of neuronal dendrites tree-like branching out to make new synaptic connection through mechanisms of dendrite morphogenesis [76]. Sprouting is a form of plastic changes in the synapses in which there is axonal synaptic reorganization to modify the efficacy of synapses [77]. LTP is the fundamental form of synaptic plasticity where synapses become strengthened and this forms the cellular basis of learning and memory [78].

Neuroplasticity is regulated by the corresponding cascade of intracellular events that translates into plastic changes. However, the plastic changes may either be adaptive, where it is related with an upsurge in function or maladaptive where it is linked with adverse consequences such as loss of function or augmented damage [79, 80]. This brings about the concept that not all plasticity effect positively on clinical status, that maladaptive plastic changes from dysregulated neuroplasticity result in an aberrant neural organization [79]. Typical example of situation where neuroplasticity becomes maladaptive can be seen in new onset of seizures after long period of cerebral trauma, where aberrant progressive plastic changes in the brain in the form of inappropriate synaptogenesis and axonal sprouting accounts for this late development. Neuroplasticity can also be seen as structural where the plastic changes involves the organization and number of synapses such as synaptogenesis, axonal sprouting and dendritic arborization, or functional where the plastic changes involves the efficacy and strength of synaptic connections such as LTP.

The basis of plastic changes that allows for neuroplasticity to become realistic depend upon factors such as neuronal excitability, which define the ability of a nerve to produce an action potential and in turn depends on the permeability, electrical and chemical state of the neuron [81]. This is then followed by adaptive changes termed plasticity, in which there are stable functional transformations that occur in specific neuronal systems as a result of specific stimuli or the combination of stimuli [82]. Furthermore, it has been revealed that effective and repeated action potentials are required from the presynaptic neuron to stimulate the postsynaptic to cause a change in the strength of an interneuron connection [83]. Cumulatively, the aforementioned process leads to biochemical changes, and anatomical adaptations which reinforce the connections between neighboring neurons, thus accounting for molecular, cellular, systems, and behavioral perspectives of explaining neuroplasticity [84].

The strength of the excitation impulse must exceed the threshold value to increase the synaptic efficacy and the stability of the connections between neurons. Nevertheless, when neurons are stimulated only with subthreshold stimuli, the overall activity of the synapse may decrease [85]. Studies conducted on unilateral lesion of the hippocampus results in the formation of new synapses (synaptogenesis) by the axons from the remaining contra-lateral hippocampal system [86]. Thus, the postsynaptic portion of a synapse continues to function properly despite the degeneration of the presynaptic region, and the surviving axons form new synapses. The fibers that form the (new) synapses are homologous to the damaged synapses, which may significantly facilitate the restoration of normal function.

5. Strategies that enhances neuroplasticity

Table 1 summarized various strategies that were found to enhance neuroplasticity and the mechanism through which modulate neuroplasticity.

Strategy	Proposed mechanism reported to modulate and promote neuroplasticity	References
Transcranial direct current stimulation (noninvasive)	Modification of neuronal membrane potentials, consequently persuading neuronal excitability which form part of the basis of neuroplasticity.	[87, 88]
Deep brain stimulation (invasive)	This by stimulating neuronal network connected to the stimulated region, the pathological neuronal network becomes altered by changes in the neurochemical components thereby inducing morphological changes in both the dendrites (dendritic arborization) and axons (axonal sprouting).	[89]
Functional Electrical Stimulation (FES noninvasive)	Hypothesized to modulate neuroplasticity through repeated generation of neurons synaptic activity that might facilitate synaptic remodeling, leading to neural reorganization.	[90]
Aerobic Exercise	Aerobic exercise is linked with surge in neurogenesis and angiogenesis, together with rise in neurotrophic molecules especially BDNF and other growth factors implicated in neurite outgrowth and synaptic plasticity	[91, 92]
Brain-derived neurotropic factor (BDNF) therapy	By binding of BDNF to its TrkB cognate receptor, two distinctive intracellular signaling pathways namely phosphatidylinositol 3-kinase (PI3K)/Akt and mitogen-activated protein kinase/extracellular-signal-regulated kinase (MAPK/ERK) becomes initiated, thereby regulating transcriptional gene activity of neurite outgrowth and neurogenesis.	[93, 94]
Statins	Proposed mechanism by which statins modulates neuroplasticity involves indirect effect through statin-mediated increase in proteins such as endothelial nitric oxide synthase (eNOS), vascular endothelial growth factor (VEGF), tissue plasminogen activator (tPA), and brain-derived neurotropic factor (BDNF) among others.	[95]
Erythropoietin (EPO) therapy	EPO and EPO receptor (EPOR) that both becomes upregulated in response to cerebral ischemia, when supplemented act to indirectly augment neurogenesis through EPO-mediated increase in the expression vascular endothelial growth factor (VEGF) and brain-derived neurotropic factor (BDNF).	[96]
Phosphodiesterase type 5 inhibitors (PDE-5 inhibitors)	PDE-5 inhibitors competitively inhibit phosphodiesterase enzymes responsible for converting cyclic guanylyl monophosphate (cGMP) back to GMP, thus fostering cGMP accumulation which has diverse cellular effect in the brain including angiogenesis, and neurogenesis which are requirements of neuroplasticity	[97]
Vascular endothelial growth factor (VEGF) therapy.	Proposed mechanism through which VEGF modulates neuroplasticity involves mediating the PI3K–AKT–nuclear factor kappa B signaling pathway; an intracellular pathway that regulate transcriptional factors involves in neurogenesis	[98, 99]

Table 1.
Various strategies that were found to enhance neuroplasticity.

6. Conclusion

Advancement in the understanding of mechanism of cerebral damage after stroke and brain neuroplasticity have continue to be a cutting-edge landmark information towards reducing human disability as a result of stroke. Strategies aimed at harnessing and augmenting neuroplasticity in complement with neurorehabilitation offers reasonable level of hope to maximize stroke recovery and diminish cerebral stroke induced neurological impairments. Although these strategies are rapidly evolving towards achieving clinical viability and success, more is needed to be done especially pertaining to outcome measures of neuroplasticity that rely on biomarkers of neuroplasticity rather than functional or behavioral outcome.

Conflict of interest

The authors declare no conflict of interest.

Author details

Mubarak Muhammad[1*] and Tasneem Muhammad Hassan[2]

1 Department of Physiology, College of Medicine, University of Ibadan, Ibadan, Nigeria

2 Department of Physiotherapy, Aminu Kano Teaching Hospital, Kano, Nigeria

*Address all correspondence to: mubarakmahmad@yahoo.com

References

[1] Kim K, Heo M, Jun H, Lee J, Jegal H, Yang Y. The relationship between stroke and quality of life in Korean adults: based on the 2010 Korean community health survey. Journal of Physical Therapy Science. 2015;27:309-312.

[2] Kuriakose D, Xiao D. Pathophysiology and treatment of stroke: present status and future perspectives. International Journal of Molecular Sciences. 2020;21:2-24.

[3] Marzolini, S, Robertson AD, Oh P, Goodman JM, Corbett D, Du X, et al. Aerobic training and mobilization early post stroke: cautions and considerations. Frontiers in Neurology. 2019;10:1-26.

[4] Strong K, Mathers C, Bonita R. Preventing stroke: saving lives around the world. Lancet Neurol. 2007;6: 182-187.

[5] Li L, Scott CA, Rothwell, PM. Trends in stroke incidence in high-income countries in the 21st century population-based study and systematic review. Stroke. 2020;51:1372-1380.

[6] Johnson W, Onuma O, Owolabi M, Sachdev S. Stroke: a global response is needed. Bulletin of the World Health Organisation. 2016;94:634-634A.

[7] Sacco RL, Kasner SE, Broderick JP, Caplan LR, Connors JJ, Culebras A, et al. An updated definition of stroke for the 21st century: A statement for healthcare professionals from the American Heart Association/ American Stroke Association. Stroke. 2013;44(7):2064-2089.

[8] Muhammad M, El-ta'alu AB, Mabrouk MI. Pathogenesis and neuro-protective agents of stroke. International Journal of Pharmaceutical Sciences and Research. 2016;7(10):3907-3916.

[9] Hendricks HT, Limbeek J, Geurts AC, Zwarts MJ. Motor recovery after stroke: a systematic review of the literature. Archives of Physical Medicine and Rehabilitation. 2002;83:1629-1637

[10] Schaechter JD. Motor rehabilitation and brain plasticity after hemiparetic stroke. Progress in Neurobiology. 2004;73:61-72.

[11] Cauraugh JH, Summers JJ. Neural plasticity and bilateral movements: a rehabilitation approach for chronic stroke patients. Progress in Neurobiology. 2005;75:309-320.

[12] Doyle S, Bennett S, Fasoli SE, McKenna KT. Interventions for sensory impairments in the upper limb after stroke (review). Cochrane Database of systematic Reviews. 2010;6:1-10.

[13] Carey LM, Matyas TA, Baum C. Effects of somatosensory impairment on participation after stroke. American Journal of Occupational Therapy. 2018;72:1-10

[14] Serrada I, Hordacre B, Hillier SL. Does sensory retraining improve sensation and sensorimotor function following stroke: a systematic review and meta-analysis. Frontiers in Neuroscience. 2019;13;1-16.

[15] Gottesman RF, Hillis AE. Predictors and assessment of cognitive dysfunction resulting from ischaemic stroke. Lancet Neurology. 2010;9(9):895-905.

[16] Lo Coco D, Lopez G, Corrao S. Cognitive impairment and stroke in elderly patients. Indian Journal of Psychological Medicine. 2016;38(3): 172-181.

[17] Yarube IU, Hassan TM, Ahmad RY, Umar LM, Musa BM, Ibrahim SMA. Association between cognition and peripheral brain-derived neurotrophic factor in a sample of normal adults in Kano, Nigeria. Nigerian Journal of Basic and Clinical Sciences. 2019;16(1):55-59.

[18] Sullivan FB. Activity-dependent factors affecting poststroke functional outcomes. Topics in Stroke Rehabilitation. 2007;8(3):31-44.

[19] Abdullahi A. Movement rehabilitation in physiotherapy after stroke: the role of constraint-induced movement therapy. Croatia: IntechOpen; 2018. p.1-19.

[20] Walter EJ, Carraretto M. The neurological and cognitive consequences of hyperthermia. Critical Care. 2016;20:199.

[21] Devan BD. Berger K, McDonald RJ. The emergent engram: A historical legacy and contemporary discovery. Frontiers in Behavioral Neuroscience. 2018;12:168.

[22] Chaudhary N, Gemmete JJ, Thompson BG, Xi, G, Pandey AS. Iron-potential therapeutic target in hemorrhagic stroke. World Neurosurgery. 2013;79(1): 4-10.

[23] Chang JJ, Emanuel BA, Mack WJ, Tsivgoulis G, Andrei V, Alexandrov AV. Matrix metalloproteinase-9: Dual role and temporal profile in intracerebral hemorrhage. Journal of Stroke and Cerebrovascular Diseases. 2014;23(10): 2498-2505.

[24] Hu X, Tao C, Gan Q, Zheng J, Li H, You C. Oxidative stress in intracerebral hemorrhage: sources, mechanisms, and therapeutic targets. Oxidative Medicine and Cellular Longevity. 2016;1-12.

[25] Warraich Z, Kleim JA. Neural plasticity: the biological substrate for neurorehabilitation. American Academy of Physical Medicine and Rehabilitation. 2010;2:S208-S219.

[26] Hara Y. Brain plasticity and rehabilitation in stroke patients. Journal of Nippon Medical School. 2015;82:4-13.

[27] Takahashi T. Novel synaptic plasticity enhancer drug to augment functional recovery with rehabilitation. Neurology. 2019;32(6):822-827.

[28] Norman SL, Wolpaw JR, Reinkensmeyer DJ. Targeting neuroplasticity to improve motor recovery after stroke. bioRxiv. https://doi.org/10.1101/2020.09.09.284620.

[29] Richards LG, Stewart KC, Woodbury ML, Senesac C, Cauraugh JH. Movement dependent stroke recovery: a systematic review and meta-analysis of TMS and FMRI evidence. Neuropsychologia. 2008;46(1):3-11.

[30] Szelenberger R, Kostka J, Saluk-Bijak J, Miller E. Pharmacological interventions and rehabilitation approach for enhancing brain self-repair and stroke recovery. Current Neuropharmacology. 2020;18:51-64.

[31] Murphy TH, Corbett D. Plasticity during stroke recovery: from synapse to behaviour. Nature Reviews Neuroscience. 2009;10(12):861-872.

[32] Levin MF, Kleim JA, Wolf SL. What do motor "recovery" and "compensation" mean in patients following stroke? Neurorehabilitation and Neural Repair. 2009;23(4):313-319.

[33] Hillis AE, Tippett DC. Stroke recovery: surprising influences and residual consequences. Advance in Medicine. 2014;378263.

[34] Liu F, Cheng X, Zhong S, Liu C, Jolkkonen J, Zhang X, et al. Communications between peripheral and the brain-resident immune system in neuronal regeneration after stroke. Frontiers in Immunology. 2020;11:1931.

[35] Canazza A, Minati L, Boffano C, Parati E, Binks S. Experimental model of brain ischaemia: a review of technique, magnetic resonance imaging, and investigational cell-based therapies. Frontier in Neurology. 2014;5(19):1-15.

[36] Zhao L, Willing A. Enhancing endogenous capacity to repair a stroke damaged brain: An evolving field for stroke research. Progress in Neurobiology. 2018;164:5-26.

[37] Clark TA, Sullender C, Jacob D, Zuo Y, Dunn AD, Jones TA. Rehabilitative training interacts with ischemia-instigated spine dynamics to promote a lasting population of new synapses in peri-infarct motor cortex. The Journal of Neuroscience. 2019;39(43):8471-8483.

[38] Friec AL, Salabert A, Davoust C, Demain B, Vieu C, Vaysse L, et al. Enhancing plasticity of the central nervous system: drugs, stem cell therapy, and neuro-implants. Neural Plasticity. 2017;1-9.

[39] Guggisberg AG, Koch PJ, Hummael FC, Buetefisch C. Brain networks and their relevance for stroke rehabilitation. Clinical Neurophysiology. 2019;130(7):1098-1124.

[40] Woodruff TM, Thundyil J, Tang S, Sobey CG, Taylor SM, Arumugam TV. Pathophysiology, treatment, and animal and cellular models of human ischemic stroke. Molecular Neurodegeneration. 2011;6(11):1-19.

[41] Stankowski JN, Gupta R. Therapeutic targets for neuroprotection in acute ischemic stroke: lost in translation? Antioxidant & Radox Signalling. 2011;14(10):1841-1851.

[42] De Keyster J, Uyttenboogaart M, Koch MW, Elting JW, Sulter G, Vroomen PC. et al. Neuroprotection in acute ischaemic stroke. Acta Neurologica Belgica. 2005;105:144-148.

[43] Peruzzotti-Jametti L, Donega M, Giusto E, Mallucci G, Marchetti B, Pluchino S. The role of the immune system in central nervous system plasticity after acute injury. Neuroscience. 2014;26(0):210-221.

[44] Matute C, Domercq M, Perez-Samartin A Ransom BR. Basic science advances for clinicians: protecting white matter from stroke injury. Stroke. 2013;44:1204-1211.

[45] Lai TW, Zhang S, Wang YT. Excitotoxicity and stroke: Identifying novel targets for neuroprotection. Progress in Neurobiology. 2014;115: 157-188.

[46] Onwuekwe IO, Ezeala-Adikaibe B. Ischemic stroke and neuroprotection. Annals of Medical & Health Sciences Research. 2012;2(2):186-190.

[47] Kalogeris T, Bao Y, Korthuis RJ. Mitochondrial reactive oxygen species: a double edged sword in ischemia/ reperfusion vs preconditioning. Redox Biology. 2014;2:702-714.

[48] Muhammad M. Tumor necrosis factor alpha: a major cytokine of brain neuroinflammation In: Cytokines. London: IntechOpen Publishers; 2020. pp.27-40. http://dx.doi.org/10.5772/ intechopen.77671

[49] Walberer M, Jantzen SU, Backes H, Rueder MA, Keuters MH, Neumaier B, et al. In-vivo detection of inflammation and neurodegeneration in the chronic phase after permanent embolic stroke in rats. Brain Research. 2014;15(81): 180-188.

[50] Majid A. Neuroprotection in stroke: past, present, and future. ISRN Neurology. 2014;1-17.

[51] Lakhan SE, Kirchgessner A, Hofer M. Inflammatory mechanisms in ischemic stroke: therapeutic approaches. Journal of Translational Medicine. 2009;7(97):1-11.

[52] Shirley R, Ord ENJ, Work LW. Oxidative stress and the use of antioxidants in stroke. Antioxidants. 2014;3:472-501.

[53] Forstermann U, Sessa WC. Nitric oxide synthases: Regulation and function. European Heart Journal. 2012;33:829-837.

[54] Velayutham M, Zweier JL. Nitric oxide signaling in biology. Messenger. 2013;2(1):1-18.

[55] Andrabi SS, Parvez S, Tabassum H. Melatonin and ischemic stroke: Mechanistic roles and action. Advances in Pharmacological Sciences, 2015;1-11.

[56] Orrenius S, Nicotera P, Zhivotovsky B. Cell death mechanisms and their implications in toxicology. Toxicological Sciences. *2011;*119(1):3-19.

[57] Broughton BRS, Reutens DC, Sobey CG. Apoptotic mechanisms after cerebral ischemia. Stroke. 2009;40: 331-339.

[58] Alvarez A, Lacalle J, Canavate ML, Alonso-Alconada D, Lara-Celador I, Alvarez F.J, et al. Cell death. A comprehensive approximation necrosis. Microscopy: Science, Technology, Applications and Education. 2010;1017-1024.

[59] Breton RR, Rodríguez JCG. Excito-toxicity and oxidative stress in acute ischemic stroke. Acute Ischemic Stroke. 2012;31-58.

[60] Kar B, Sivamani S. Apoptosis: basic concepts, mechanisms and clinical implications. International Journal of Pharmaceutical Sciences and Research. 2015;6(3):940-950.

[61] Keep RF, Hua Y, Xi G. Intracerebral haemorrhage: mechanisms of injury and therapeutic targets. Lancet Neurol. 2012;11(8):1-25.

[62] Aronowski J, Zhao X. Molecular pathophysiology of cerebral hemorrhage: Secondary brain injury. Stroke. 2011;42(6):1781-1786.

[63] Chang JJ, Emanuel BA, Mack WJ, Tsivgoulis G, Andrei V, Alexandrov AV. Matrix Metalloproteinase-9: Dual role and temporal profile in intracerebral hemorrhage. Journal of Stroke and Cerebrovascular Diseases. 2014;23(10):2498-2505.

[64] Schmidt S, Gull S, Herrmann K, Boehme M, Irintchev A, Urbach A, et al. Experience-dependent structural plasticity in the adult brain: How the learning brain grows. bioRxiv. 2020;1-52.

[65] Dobkin BH. Rehabilitation after stroke. The New England Journal of Medicine. 2005;352:1677-1684.

[66] Rossini PM, Calautti C, Pauri F, Baron J. Post-stroke plastic reorganisation in the adult brain. The Lancet Neurology. 2003;2:493-502.

[67] Stewart, JC, Cramer SC. Genetic variation and neuroplasticity: role in rehabilitation after stroke. Journal of Neurologic Physical Therapy. 2017;41(3):S17-S23.

[68] Alawieh A, Andersen M, Adkins DL, Tomlinson S. Acute complement inhibition potentiates neurorehabilitation and enhances tPA-mediated neuroprotection. The Journal of Neuroscience. 2018;38(29):6527-6545.

[69] Dąbrowski J, Czajka A, Zielińska-Turek J, Jaroszyński J, Furtak-Niczyporuk M, Mela A, et al. Brain functional reserve in the context of neuroplasticity after stroke. Neural Plasticity. 2019;1-9.

[70] Teasell R, Bayana NA, Bitensky J. Plasticity and reorganization of the brain post stroke. Topic in Stroke Rehabilitation. 2005;12(3):11-26.

[71] Lai KO, Wong ASL, Cheung MC, Xu P, Liang Z, Lok KC, et al. TrkB phosphorylation by Cdk5 is required for activity-dependent structural

plasticity and spatial memory. Nature Neuroscience. 2012;15:1506-1515.

[72] Hogan MK, Hamilton GF, Horner PJ. Neural stimulation and molecular mechanisms of plasticity and regeneration: a review. Frontiers in Cellular Neuroscience. 2020;14;271.

[73] Nudo RJ, Friel KM. Cortical plasticity after stroke: implications for rehabilitation. Rev Neurol. 1999;155(9):713-717.

[74] Ming G, Song H. Adult neurogenesis in the mammalian brain: significant answers and significant questions. Neuron. 2011;70(4):687-702.

[75] Hong J, Park M. Understanding synaptogenesis and functional connectome in *C. elegans* by imaging technology. Front. Synaptic Neurosci. 2016;8:18.

[76] Jan Y, Jan LY. Branching out: mechanisms of dendritic arborization. Nature Reviews Neuroscience. 2010; 11(5):316-328.

[77] Boele H, Koekkoek SKE, De Zeeuw CI, Ruigrok JH. Axonal sprouting and formation of terminals in the adult cerebellum during associative motor learning. The Journal of Neuroscience. 2013;33(45):17897-17907.

[78] Bliss T, Collingridge G, Morris P. Synaptic plasticity in the hippocampus In: The hippocampus book. London: Oxford University Press; 2007. Pp343-474.

[79] Cramer SC, Sur M, Dobkin BH, O'Brien CO, Sanger TD, Trojanowski JQ, et al. Harnessing neuroplasticity for clinical applications. Brain. 2011;134(6): 1591-1609.

[80] Carey L, Walsh A, Adikari A, Goodin P, Alahakoon D, De Silva D, et al. Finding the intersection of neuroplasticity, stroke recovery, and learning: scope and contributions to stroke rehabilitation. Neural Plasticity. 2019;1-15.

[81] Nudo RJ, Plautz EJ, Frost SB. Role of adaptive plasticity in recovery of function after damage to motor cortex (2001). Muscle Nerve. 2001;24:1000-1019.

[82] Sanchez-Mendoza EH, Hermann DM. Correlates of post-stroke brain plasticity, relationship to pathophysiological settings and implications for human proof-of-concept studies. Frontiers in Cellular Neuroscience. 2016;10:196.

[83] Bohotin CR, Badescu M, Popescu DN, Bohotin V. Motor cortex plasticity: from physiology to clinical neurology. Romanian Journal of Physiology. 2004;41:99-108.

[84] Nudo RJ. Functional and structural plasticity in motor cortex: implications for stroke recovery. Physical Medicine and Rehabilitation Clinics of North America. 2003;14:S57-S76.

[85] Poirazi P, Mel BW. Impact of active dendrites and structural plasticity on the memory capacity of neural tissue, Neuron. 2001;29(3):779-796.

[86] Kania BF, Wronska D, Zięba D. Introduction to neural plasticity mechanism. Journal of Behavioral and Brain Science. 2017;7:41-49.

[87] Fregni F, Pascual-Leone A. Technology insight: noninvasive brain stimulation in neurology-perspectives on the therapeutic potential of rTMS and tDCS. Nature Clinical Practice Neurology. 2007;3:383-393.

[88] Reis J, Schambra HM, Cohen LG, Buch ER, Fritsch B, Zarahn E, et al. Noninvasive cortical stimulation enhances motor skill acquisition over multiple days through an effect on consolidation. Proc Natl Acad Sci. 2009;106:1590-1595.

[89] Johnson MD, Miocinovic S, McIntyre CC, Vitek JL. Mechanisms and targets of deep brain stimulation in movement disorders. Neurotherapeutics. 2008;5:294-308.

[90] Chen J, Zacharek A, Li A, Zhang C, Ding J, Roberts C, et al. Vascular endothelial growth factor mediates atorvastatin-induced mammalian achaete-scute homologue-1 gene expression and neuronal differentiation after stroke in retired breeder rats. Neuroscience. 2006;141(2):737-744.

[91] Rhyu IJ, Bytheway JA, Kohler SJ, Lange H, Lee KJ, Boklewski J, et al. Effects of aerobic exercise training on cognitive function and cortical vascularity in monkeys. Neuroscience. 2010;167:1239-1248.

[92] Hassan TM, Yarube IU. Peripheral brain-derived neurotrophic factor is reduced in stroke survivors with cognitive impairment. Pathophysiology. 2018;25(4):405-410.

[93] Tanaka K, Okugawa Y, Toiyama Y, Inoue Y, Saigusa S, Kawamura M, et al. Brain-derived neurotrophic factor (BDNF)-induced tropomyosin-related kinase b (TRK B) signaling is a potential therapeutic target for peritoneal carcinomatosis arising from colorectal cancer. PLoS One. 9(5):e96410.

[94] Kuipers SD, Trentani A, Tiron A, Mao X, Kuhl D, Bramham CR. BDNF-induced LTP is associated with rapid Arc/Arg3.1-dependent enhancement in adult hippocampal neurogenesis. Scientific Reports. 2016;6:21222.

[95] Walter DH, Rittig K, Bahlmann FH, Kirchmair R, Silver M, Murayama T, et al. Statin therapy accelerates reendothelialization: a novel effect involving mobilization and incorporation of bone marrow-derived endothelial progenitor cells. Circulation. 2002;105(25):3017-3024.

[96] Keswani SC, Buldanlioglu U, Fischer A, Reed N, Michelle P, Liang H, et al. A novel endogenous erythropoietin mediated pathway prevents axonal degeneration. Ann Neurol. 2004;56:815-826.

[97] Reierson GW, Guo S, Mastronardi C, Licinio J, Wong M (2011); cGMP signaling, phosphodiesterases and major depressive disorder. Current Neuropharmacology. 2011;9: doi : 10.2174/157015911798376271

[98] Zhang ZG, Zhang L, Jiang Q, Zhang R, Powers C, Bruggen NV, et al. VEGF enhances angiogenesis and promotes blood-brain barrier leakage in the ischemic brain. Journal of Clinical Investigation. 2000;106(7):829-838.

[99] Malykhina AP, Lei Q, Erickson CS, Epstein MI, Saban MR, Davis CA, et al. VEGF induces sensory and motor peripheral plasticity, alters bladder function, and promotes visceral sensitivity. BMC Physiology. 2012;12(15):1-21.

Chapter 4

Movement-Related Cortical Potential Associated with Jaw-Biting Movement in the Patients with Oral Cancer after the Surgery

Ichiro Nakajima, Mitsuyo Shinohara and Hiroiku Ohba

Abstract

Oral cancer is first treated with surgery for the patients. In most cases, it becomes difficult for these patients to perform smooth jaw movements postoperatively, causing masticatory dysfunctions, due to the mandible excision including muscles and peripheral nerves. However, it is still unknown whether the surgery affects the brain function for jaw movement in the patients. In this study, therefore, we investigated a significance of the movement-related cortical potential (MRCP) for jaw movements in the patients after the cancer surgery, to clarify the motor preparation process in the brain, as compared with healthy subjects. Eight normal subjects and seven patients with oral cancers were enrolled in the study. Experiment 1: The normal subjects were instructed to perform jaw-biting movement and hand movement, respectively. The MRCPs appeared bilaterally over the scalp approximately 1 to 2 s before the onset of muscle discharge in both movements. Experiment 2: The MRCPs appeared preoperatively in the jaw biting movement in all patients. However, the amplitudes of the MRCP decreased significantly after than before the surgery ($p < 0.05$). Our data indicated the dysfunction of the motor preparation process for jaw movements in the patient after the surgery, suggesting impairment of feed-forward system in the maxillofacial area.

Keywords: oral cancer, mandibulectomy, movement-related cortical potential, jaw movement, feed-forward system

1. Introduction

According to the World Health Organization, there are an estimated 657,000 new cases of oral cavity and pharyngeal cancers each year, and more than 330,000 death [1]. In high-risk countries such as Sri Lanka, India, Pakistan and Bangladesh, oral cancer is the most common cancer in men, and may contribute up to 25% of all new cases of cancer [2].

Oral cancers can occur on the lip or in the oral cavity, nasopharynx, and pharynx. It belongs to a larger group of cancers called head and neck cancers. Usually, oral cancer is first treated with surgery. In most cases, surgery is followed by radiation therapy and chemotherapy [3–5].

During the jaw bone excision in oral cancer surgery, we commonly find that the adjacent masticatory muscle, tendon tissue, and peripheral nerves are known to be included in the disease field. However, oral cancer patients who underwent resection surgery can suffer from eating and swallowing dysfunctions, owing to the masticatory muscle's excision, even when bone grafts obtain a morphological recovery [6].

In the postoperative period, most patients complain of immobility at the onset of jaw movement.

We predict that one cause is a disorder associated with the preparatory stage of the onset of voluntary jaw movement in the patients, leading to difficulty masticating and swallowing.

Jaw motor dysfunctions may be related either to the mandibular bone and teeth' defects or some brain functions' modulation after surgery.

Movement-related cortical potential (MRCP) is a slow negative potential in an electroencephalographic recording that occurs about 2 s before voluntary body movement production in humans [7].

In general, MRCP consists of two main components: Bereitschaftspotential (BP) and negative slope. From extensive studies, the current model is that BP starts first in the SMA, including the pre-SMA, and then shortly after that in the lateral premotor cortices bilaterally [8, 9]. About 400 msec before the movement onset, NS' starts in M1 and the premotor cortex mainly. MRCPs are also generated from the cerebellum as subcortical structures [10].

This potential is well known to reflect the cortical processes involved in movement planning and movement preparation preceding voluntary limb and maxillofacial movements [11–16].

In this study, we observed changes in MRCP waveform components associated with jaw-biting movements in oral cancer patients before and after surgical excision of their lesions. The purpose of this study was to investigate whether brain activities, as measured by MRCP recordings, are affected by the loss of neurological tissue in patients with oral cancer after the surgery.

In this chapter, thus, we discuss the clinical importance of application of MRCP recording in the field of maxillofacial surgery.

2. Material and methods

2.1 Research design

We performed two experiments in this study, described below.

Experiment 1: The MRCP waveforms for jaw movement were compared with those for hand movement in healthy subjects, which served as the control, to confirm MRCP components' characteristics (BP and NS') preceding onset of jaw muscle activities.

Experiment 2: The MRCP waveforms for jaw movement were compared between the preoperative and postoperative periods in the oral cancer patients to confirm MRCP components' changing (BP and NS') preceding onset of jaw muscle activities.

2.2 Informed consent

We obtained informed consent was obtained from all participants after explaining the procedure in detail. The Ethical Committee approved the protocol of Nihon University School of Dentistry.

2.3 Experiment 1

2.3.1 Subjects

Eight unaffected subjects, four males and four females were enrolled in this experiment. Each had a complete set of natural maxillary teeth, instead of mandibular teeth. The average age was between 25 and 32 years.

2.3.2 Motor tasks

2.3.2.1 Jaw-biting task

The subjects were instructed to bite softly on a 3-cm-thick plastic block between their upper and lower molars on their habitual side (left or right) at their own pace.

2.3.2.2 Wrist dorsiflexion task (control movement task)

The subjects were instructed to put their hands on the chair's arms and dorsiflex their right wrist.

The above movement was carried out 50 times rapidly, and three trials were carried out with a break between them. The MRCP waveform data obtained from both groups for the above movement tasks were compared with those obtained for the right wrist dorsiflexion movement task as the control task.

The subjects sat in an armchair with the FH plane nearly parallel to the floor in a relaxed manner. They were instructed not to move their jaw and tongue before doing each movement.

The subjects performed the movements 30–50 times rapidly and carried out the following two tasks with a break between them. The subjects were instructed to gaze at a mark positioned 1.5 m ahead at eye level lightly. The subjects were instructed not to blink for 6 s before beginning the specified movement and to relax, and then to perform the following movement tasks.

2.3.3 Recording conditions

2.3.3.1 Electroencephalographic (EEG) recording

For electroencephalographic recording, Ag/AgCl electrodes were set at three sites, C3, Cz, and C4, stipulated under the international 10–20 system (**Figure 1**). The right and left earlobes (A1 and A2) were selected as the reference electrodes.

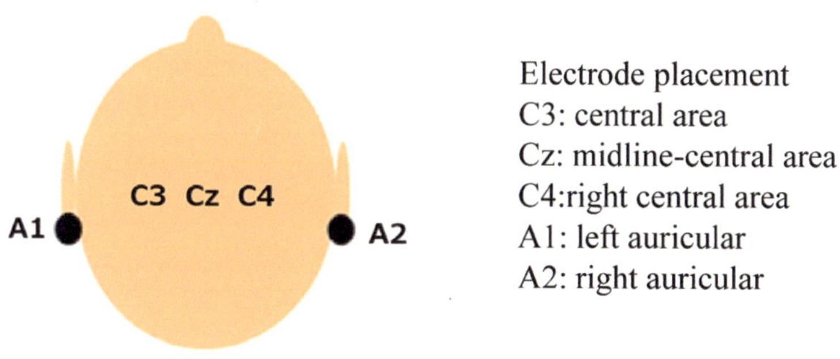

Figure 1.
Electroencephalographic recoding areas.

Electroencephalography was performed using a monopolar lead with a time constant of 5.0 s, a 100 Hz high-frequency filter, and electric resistance of 5 kΩ or less.

2.3.3.2 Electrooculograms (EOG) recording

Electrooculograms were recoded from electrodes above and below the right eye, to enable us to monitor for eye movement and blink artifacts.

2.3.3.3 Electromyogram (EMG) recording

Surface electromyograms that served as a trigger for brain wave averaging were recorded from agonists shown below:

Jaw-biting task: Superficial part of the right masseter muscle.

Wrist movement task: Short radial extensor muscle of the right wrist (control movement task).

Surface electromyograms were recorded from each muscle group using a bipolar lead. The ocular movement was recorded during all the movement tasks, and electroencephalographic data contaminated with ocular movement artifacts were excluded from the analysis.

2.3.4 Data analysis conditions

Signals obtained from electroencephalography, ocular movement, and muscle activity were amplified with an evoked potential measuring system (Neuropak-MEB-2200, Nihon Kohden Corporation). Electroencephalographic data without artifacts taken 50 times with the point of muscle discharge of the suprahyoid muscle group exceeding 2 μV chosen as the threshold for the trigger were averaged.

2.4 Experiment 2

2.4.1 Patients

Seven oral cancer patients in whom the large area of the mandibular bone was excised during tumor resection were enrolled in this study (**Table 1**).

The patients ranged in age from 17 to 70 at the time of surgery. Additionally, they had upper and lower molars on the non-operative side (unaffected side). Their operation procedures were either segmental mandibulectomy or hemi-mandibulectomy (**Figure 2**).

2.4.2 Motor task

The patients were instructed to carry out the jaw-biting task on the unaffected side. They were instructed to bite softly on a 3-cm-thick plastic block between the upper and lower molars on the unaffected side at their own pace before undergoing tumor resection. The patients carried out the same specified movement postoperatively. One of the patients was instructed to perform the right wrist dorsiflexion movement postoperatively.

2.4.3 Recording conditions

Experiment 2 was carried out according to the recording conditions of Experiment 1. Electroencephalography was done on five patients, according to the

Case	Age	Diagnosis	Surgical procedure	Recorded month of MRCP before surgery	Recorded month of MRCP after operation
1	36Y	Fibrosarcoma	Left hemi-mandiblectomy	1 month within	12 month
2	60Y	Squamous cell carcinoma	Right segmental mandiblectomy		
3	63Y	Squamous cell carcinoma	Right segmental mandiblectomy		
4	70Y	Squamous cell carcinoma	Left segmental mandiblectomy		
5	17Y	Ameloblastoma	Left segmental mandiblectomy		
6	18Y	Ameloblastoma	Left segmental mandiblectomy		
7	70Y	Ameloblastoma	Left segmental mandiblectomy		

Table 1.
The list of patients participated in the research.

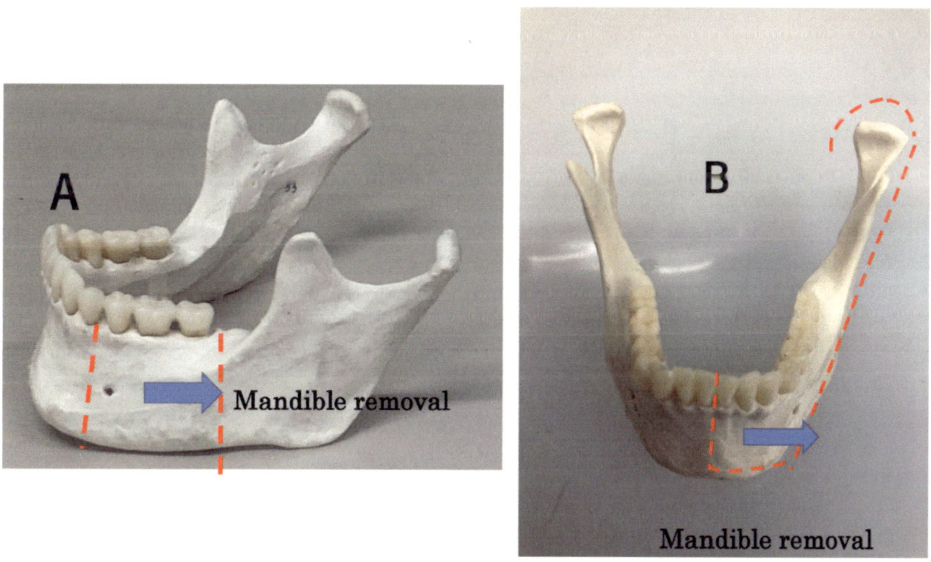

Figure 2.
Surgical areas of mandibulectomy in the patients. (A) Segmental mandibulectomy is a surgical procedure in which the mandible is partially resected continuously from the alveolus to the inferior border of the mandible, resulting in the disruption of its continuity. (B) Hemi-mandibulectomy is a surgical procedure involving resection of the hemi-mandible including its condyle. Between 1 to 3 years after the surgery, the mandibular bone was reconstructed with bone grafts from the ilium in these patients. The patients who participated in this study also complained of difficulty in jaw movement and mastication after surgery.

international 10–20 method, by placing sensors over the vertex (CZ). For electromyography, recordings were made using bipolar leads placed over the left and right masseter muscles. Simultaneously, electrooculography was done to monitor a mixture of eye-blinking potential with theelectroencephalogram.

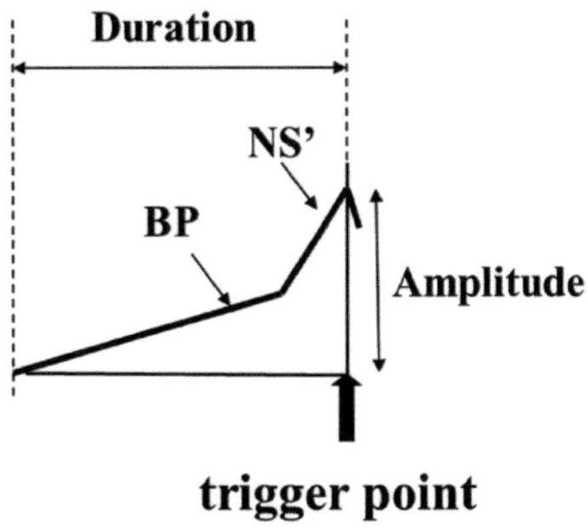

Figure 3.
MRCP components. Bereitschaftspotential (BP): BP is a gradually increasing, bilaterally widespread surface negativity with the maximum over the midline vertex region regardless of the site of movement onset. Negative slope (NS'): NS' is a much steeper slope starting about 400 ms before the movement onset, and is characterized by a more localized negativity over the central and vertex regions contralateral to the movement. Trigger point indicates the onet of EMG for EEG averaging.

2.4.4 Measurement parameters

The measurement parameters for each movement task were maximum amplitude and duration determined by setting the baseline and rise time of MRCP under the following conditions (**Figure 3**). The average potential of a segment between 4 and 3 s before the start of muscle discharge was calculated to set the baseline.

2.4.5 MRCP duration, maximum MRCP amplitude

2.4.5.1 Maximum MRCP amplitude

The difference in potential between peak MRCP amplitude before muscle discharge initiation and the baseline (NS' amplitude) was determined as the maximum MRCP amplitude.

2.4.5.2 MRCP duration

The difference in time between the starting point of the rise in MRCP from the baseline and muscle discharge initiation was determined as the MRCP duration.

2.5 Statistical analysis

Differences of the amplitudes in the two different motor tasks were analyzed with a two-factor (recording site - task), repeated-measures analysis of variance (ANOVA). Statistical significance was considered when $P < 0.05$. A paired t-test was also used to compare the measured values (amplitude/duration) of two groups.

3. Results

3.1 Experiment 1

3.1.1 MRCP waveform components in tasks

MRCP waveforms obtained in jaw-biting and wrist motor tasks were shown in Fi.4A and **Figure 4B**, respectively. Negative slow potential appeared bilaterally

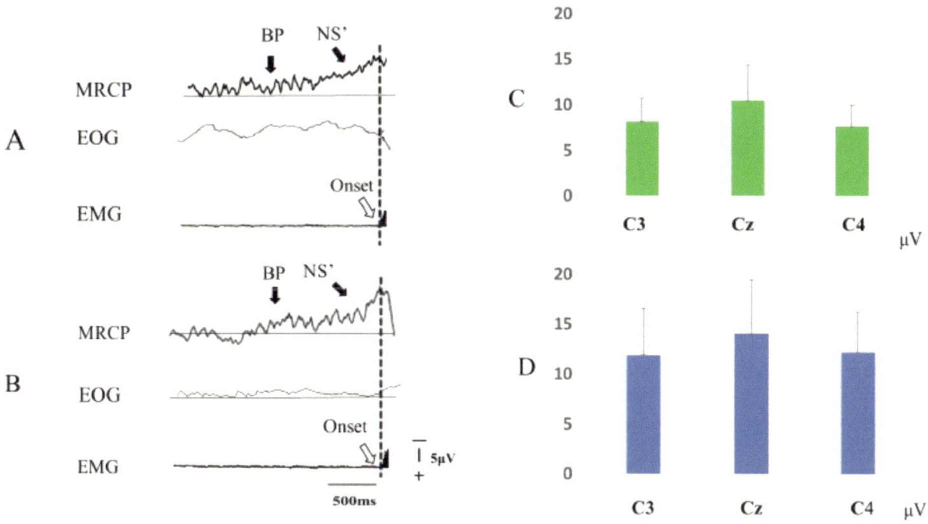

Figure 4.
Comparison of cortical potentials in jaw biting and wrist extension tasks in healthy subjects. (A) The grand averaged MRCPs for wrist extension task over all eight subjects. (B) The grand averaged MRCPs for jaw-biting task over all eight subjects. Both waveforms were obtained from vertex (Cz). The BP component and NS component can be confirmed in each waveform."Onset" indicates onset of muscle discharge. Panel (C) and (D) indicate mean peak amplitudes values of the MRCPs for wrist extension and for jaw-biting tasks, respectively.

from electrodes at five sites on the scalp 1 or 2 s before muscle discharge initiation in each task. Each slow negative potential was divided into the BP component, which was of long duration centering on Cz, and the rapid NS' component appearing approximately 500 msec before muscle discharge initiation.

3.1.2 Maximum peak amplitudes (NS' amplitude) in each task

3.1.2.1 Mean peak amplitudes of MRCPs

Maximum amplitudes in the wrist dorsiflexion movement task were 8.18 ± 2.46 μV, 10.48 ± 3.84 μV, 7.64 ± 2.41 μV, at C3, Cz, and C4, respectively (**Figure 4C**). On the other hand, the MRCP amplitudes in the jaw-biting task were 12.80 ± 5.22 μV, 15.09 ± 5.98 μV, 12.91 ± 4.32 μV at C3, Cz, and C4, respectively (**Figure 4D**).

In a two-way analysis of variance with each task and each recording site as two factors, a significant difference in MRCP amplitude was observed when the task was considered as the main effect [$F(1,42) = 11.8$, $p < 0.01$]. No significant difference was also observed between the recording sites [$F(2,42) = 1.78$, $p > 0.05$]. No interaction was observed [$F(2,42) = 0.07$, $p > 0.05$]. These results indicate that MRCP amplitude varied with the task. The MRCP amplitude was significantly higher in the jaw-biting task than in the wrist dorsiflexion movement task ($p < 0.01$). The MRCP recorded at Cz tended to be a large amplitude in both tasks, and was clear to distinguish two main components (BP and NS').

3.1.3 MRCP duration in tasks

The MRCP durations were 1,704.37 ± 743.03 ms in the wrist dorsiflexion movement task, 2,180.00 ± 196.81 ms in the jaw-biting task. No significant difference was observed in MRCP duration between the movement tasks (paired test, $p > 0.05$).

3.2 Experiment 2

3.2.1 Changes in MRCP waveform components in oral cancer patients

Regarding MRCP waveforms associated with the preoperative jaw-biting, negative potential appeared bilaterally 1.5 s before muscle discharge initiation. The precipitous negative potential appeared 500 msec before muscle discharge initiation in all five patients. Preoperative negative potential waveforms consisted of BP and NS' components.

However, BP and NS' appearance became unclear in four of all seven patients after the surgery. In only one case, BP and NS' components were observed at both pre and postoperative periods.

Figure 5 shows a typical case with reduced waveform components. In this case, the BP component was confirmed 1.5 seconds before the onset of masseter muscle discharge, as in healthy adults, before the surgery (**Figure 5A**).

However, after surgery, the amplitudes of MRCPs were decreased so much that the BP component could not be confirmed (**Figure 5B**). The NS' component was barely confirmed in the waveforms. In addition, the BP and NS components could be observed in the MRCP waveforms when this patient was requested to perform wrist movement during the postoperative period (**Figure 5C**).

The MRCP amplitudes in the jaw-biting task at pre-and postoperative periods were 6.80 ± 1.442 μV and 4.02 ± 2.33 μV, respectively (**Figure 6**). The amplitudes

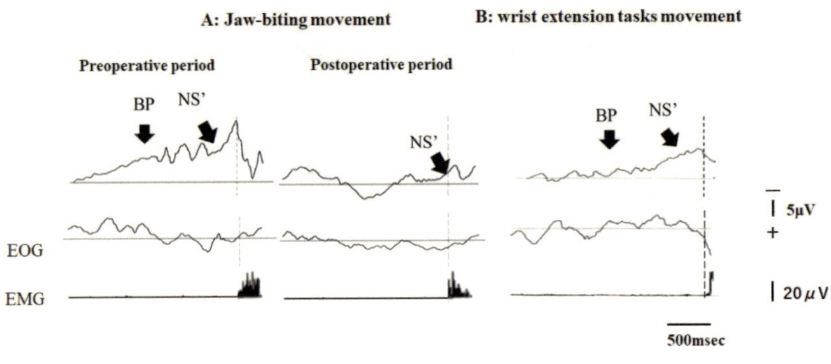

Figure 5.
A typical change in pre and postoperative MRCP waveforms in same patient. (A) Comparison between the MRCPs for jaw-biting in a patient at preoperative and postoperative periods. (B) The MRCPs for wrist extension in same patient at postoperative period.

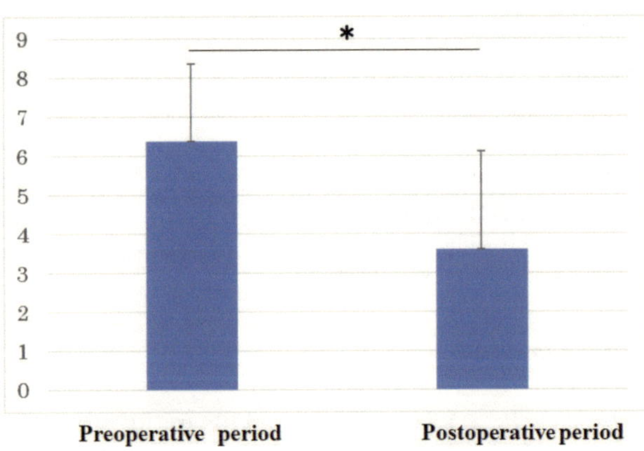

Figure 6.
*Comparison of mean peak amplitude values before and after surgery. Paired t test: * indicates significant difference at p < 0.05.*

of the MRCP decreased significantly in the patients after than before the surgery ($p < 0.05$). Thus, it was clear that the cerebral potentials associated with jaw movement were affected in the patient after surgery.

4. Discussion

4.1 Experiment 1

Recently, the epidemiological data suggest a positive correlation between masticatory functions and cognitive functions in the elderly people [17, 18]. Thus, we feel it is essential to study the relationship between masticatory muscle movement and brain function in humans from many points of view.

That local cerebral function is related to various movements' execution has been documented recently using cerebral function imaging methods such as fMRI and positron-emission tomography [19–21]. These techniques have excellent spatial resolution and various applications, such as function localization in the cerebral cortex. These techniques can be used in jaw movement studies.

Because these methods' temporal resolutions, which are on the order of tens to hundred milliseconds, are low, it is impossible to record brain activity on a time series plot: before, at the start of, and during a movement. The methods provide information on brain activity during a movement but cannot clarify brain activity during a movement's preparation.

On the other hand, an MRCP is recorded from the scalp before voluntary movements. Hence, an index reflecting changes in brain potential, but MRCP reflects the supplementary motor area and sensorimotor cortex's activities appear before the start of voluntary movements. Its temporal resolution is on the order of 1/1000 seconds, allowing very high-resolution examination and enabling continuous monitoring of central nerve activity accompanying the movement.

Shibasaki (1980) divided MRCP recorded from the scalp into two components: the BP and the NS' components [8]. The BP component initiates approximately 1,000 to 2,000 ms before starting a movement with a mild gradient showing maximum amplitude. In contrast, the NS' component initiates at approximately 500 ms after the start of a movement with a steep gradient showing maximum amplitude on the side contralateral to the movement side. The NS' measurement is used clinically to diagnose movement disorder of central origin.

Therefore, we conducted this study by focusing on changes in the BP and NS' components of MRCP waveforms associated with jaw movements.

In this study, the BP component appeared bilaterally from the scalp 1 to 1.5 s ahead of muscle discharge initiation in the jaw-biting task, similarly to the wrist dorsiflexion movement task. However, the NS' amplitude tended to be higher in the jaw-biting task than in the wrist dorsiflexion movement task. Our data demonstrated the difference between jaw and limb movements in the peak amplitudes of MRCPs.

MRCP amplitude has been shown to have a more significant increase, particularly in the movement task coordinating the bilateral upper limbs than in the unilateral upper-limb movement task [21, 22]. This phenomenon is considered attributable to increased activity in the supplementary motor area. The difference in MRCP amplitude between the jaw movement and wrist movement tasks was considered to indicate the brain's more extensive activation in the jaw movement's preparatory process involving the bilateral muscles' coordination than for the unilateral wrist dorsiflexion movement.

Our results indicate that BP and NS' components can be considered a useful index for the jaw-biting movement's motor preparation, similar to limb movements. In later experiments, we will investigate the relationship between these components and the brain's active sites for jaw-biting movement in patients with oral cancer.

4.2 Experiment 2

In Experiment 1, our data indicated that BP and NS' were a useful index of MRCPs, suggesting the more complicated preparatory process for the jaw movement than for the limb movement. In Experiment 2, the influence of excision of the unilateral masticatory muscle in oral tumor resection on the BP and NS' components of the MRCPs for the jaw-biting was examined in the patients.

Chewing aims to crush, triturate, and mix food with saliva so that that food can be transported by deglutition down the digestive canal [23]. The masticatory muscle is an agonist of an actual jaw movement and a sensory organ involved in sensing the mandible position through stretch receptors [24].

Therefore, in this experiment, we predicted that a loss of oral sensory receptors such as muscle spindles due to oral cancer removal might affect brain function.

In this study, MRCP consisting of the BP and NS' components appeared initial onset of master muscle activities in all five patients at the preoperative period.

However, these MRCPs decreased in four of five subjects during the postoperative period in the patients. In particular, our results indicated the BP components more remarkably decreased.

The cerebellum receives signals from receptors on various parts of the body and the cerebrum [25, 26].

The cerebellum transmits movement-controlling signals based on the above signals to the medulla's vestibular nuclei oblongata and cerebellar nuclei. Various sensory input types are distributed fragmentarily in intracortical adjacent areas, and Purkinje cells, the only projection neurons, potentially integrate various sensory information at a single-cell level. It is considered that the plasticity of parallel fiber synapses determined by the interaction of two types of excitatory input received by Purkinje cells, climbing fiber inputs and parallel fiber inputs, plays an important role in sensory information processing and motor learning.

It is well known that the cerebellum drives MRCP generation. Ikeda et al. [27] have determined the cortical source of MRCPs by directly recording cortical potentials through chronically implanted subdural electrodes in patients with epilepsy as part of the presurgical evaluation [10]. Sasaki et al. [28] reported that in monkeys trained to perform spontaneous hand movements, MRCPs recorded before the movement's initiation disappear following cerebellum resection [28]. Ikeda et al. [27] reported that in cerebellar infarction patients performing an upper-limb movement task, no MRCP was recorded [27].

Naito et al. demonstrated that sensorimotor inputs from muscle spindles and tendon receptors through afferent nerves in humans are transmitted to the primary sensorimotor cortices, supplementary motor area (SMA) and cingulate motor area (CMA), by using emission tomography (PET) [29].

In this study, therefore, our data may suggest that the marked postoperative reduction in MRCP amplitude in the oral cancer patients was attributable to decreased activities of the cerebellum before the initiation of voluntary movements, leading to dysfunction of the feed-forward system of jaw movements.

As a further study, we will investigate the effect of postoperative rehabilitation for cancer patients using not only MRCP but f MRI analysis, in order to establish the effective dysphagia treatment.

5. Conclusion

The above findings support our hypothesis that the removal procedure of the oral cancer including masticatory muscles (muscle spindle, tendon receptor) may affect this information processing of brain for smooth jaw movement. We concluded that it is necessary to diagnose brain functions for the oral cancer patients before and after the surgery for postoperative rehabilitation of a jaw motor disorder.

Conflict of interest

The authors declare no conflict of interest.

Thanks

Authors would like to thank Prof. Mitsuhiko Matsumoto for his warm research support sincerely.

Author details

Ichiro Nakajima[1*], Mitsuyo Shinohara[2] and Hiroiku Ohba[3]

1 Department of Community Dentistry, Nihon University School of Dentistry, Tokyo, Japan

2 Department of Oral and Maxillofacial Surgery, Faculty of Medicine, Juntendo University, Tokyo, Japan

3 Department of Dental and Oral Surgery, Tokyo Saiseikai Central Hospital, Tokyo, Japan

*Address all correspondence to: nakajima.ichirou@nihon-u.ac.jp

References

[1] WHO. Oral cancer. [Internet]. 2020. Available from https://www.who.int/cancer/prevention/diagnosis-screening/oral-cancer/en/ [Accessed 2020-11-22]

[2] Warnakulasuriya S : Global epidemiology of oral and oropharyngeal cancer. Oral Oncology. 2009;45(4-5):309-16.DOI:10.1016/j.oraloncology.2008.06.002

[3] Montero H P, Patel G S: Cancer of the oral cavity. Surgical Oncology Clinics of North America. 2015; 24(3): 491-508. DOI: 10.1016/j.soc.2015.03.006.

[4] Moroi H. H., Okimoto, K. , Terada, Y. :The effect of an oral prosthesis on the quality of life for head and neck cancer patients. Journal of Oral Rehabilitation. 1999; 26:265-273. DOI.org/10.1046/j.1365-2842.1999.00379.x

[5] Specht L.: Oral complications in the head and neck radiation patient. Support Care Cancer. 2002; 10(1):36-39. DOI: 10.1007/s005200100283

[6] Namaki S., Matsumoto M., Ohba H., Tanaka H., Koshikawa N. Shinohara M.:Masticatory efficiency before and after surgery in oral cancer patients: comparative study of glossectomy, marginal mandibulectomy and segmental mandibulectomy. Journal of Oral Science. 2004; 46 (2):113-117. DOI:10.2334 / josnusd.46.113

[7] Kornhuber H. H., Deecke L.: Brain potential changes in voluntary and passive movements in humans: readiness potential and reafferent potentials. Pflügers Archiv-European Journal of Physiology. 2016; 468 (7): 1115-24. DOI:10.1007/s00424-016-1852-3

[8] Shibasaki H., Barret G., Halliday E. : Components of the movement-related cortical potential and their scalp topography. Electroencephalography Clinical Neurophysiology. 2016; 49 (3-4): 213-226. DOI:.org/10.1016/0013-4694(80)90216-3

[9] Shibasaki H. :Human brain mapping: Hemodynamic response and electrophysiology. Clinical Neurophysiology. 2008; 119 (4): 731-743. DOI: 10.1016/j.clinph.2007.10.026

[10] Ikeda A., Luders H.O., Burgess R. C., Shibasaki H. : Movement-related potentials recorded from supplementary motor area and primary motor area. Role of supplementary motor area in voluntary movements. Brain. 1992; 115 (4):1017-1043. DOI: 10.1093/brain/115.4.1017

[11] Nakajima I., Tanaka Y., Uchida A., Sakai T., Akasaka M., Mori A., Sumino R.: Cortical potentials associated with voluntary biting movement in humans. Neuroscience Research. 1991; 10 (4): 285-289. DOI: 10.1016/0168-0102(91)90085-d

[12] Nakajima I., Miyauchi M., Minowa K., Akasaka M., Uchida A. : Contingentnegativevariationsassociated with jaw opening in humans. Somatosensory and motor research. 1994; 11 (2): 149-152. DOI: org/10.3109/08990229409028867

[13] Huckabee, M. L., Deecke L., Cannito M.P., Gould H. J. , Mayr, W. : Cortical control mechanisms in volitional swallowing: the Bereitschaftspotential. Brain Topography. 2003; 16 (1) :3-17. DOI: 10.1023/a:1025671914949.

[14] Nonaka T., Yoshida M., Yamaguchi T., Uchida, A., Ohba, H., Oka S., Nakajima I. :Contingent negative variations associated with command swallowing in humans. Clinical Neurophysiology. 2009; 120 (10) :1845-1851. DOI: 10.1016/j.clinph.2009.06.029

[15] Satow T., Ikeda A., Yamamoto J., Begum T., Thuy D. H. D., Matsuhashi M., Mima T.; Nagamine T., Baba K., Mihara T., Inoue Y., Miyamoto S., Hashimoto N., Shibasaki H. :Role of primary sensorimotor cortex and supplementary motor area in volitional swallowing: a movement-related cortical potential study. American journal of physiology. Gastrointestinal and Liver Physiology. 2004; 287 (2) : 459-470. DOI: 10.1152/ajpgi.00323.2003

[16] Yoshida K., Kaji R., Kohara N., Murase N., Ikeda A,, Shibasaki H., Iizuka T. :Movement-related cortical potentials before jaw excursions in oromandibular dystonia. Movement Disorders. 2003; 18(1):94-100. DOI: 10.1002/mds.10296

[17] Miura H., Yamasaki K., Kariyasu M., Miura K., Sumi Y. :Relationship between cognitive function and mastication in elderly females. Journal of Oral Rehabilitation. 2003; 30(8):808-811. DOI: 10.1046/j.1365-2842.2003.01124.x

[18] Takeshita H., Ikebe K., Gondo Y., Inagaki H., Masui Y., Inomata C., Mihara Y., Uota M., Matsuda K., Kamide K., Takahashi R., Arai Y. Maeda Y. :Association of Occlusal Force with Cognition in Independent Older Japanese People. JDR Clinical and Translational Research. 2016; 1(1):69-76. DOI: 10.1177/2380084416636604.

[19] Deiber P.M., Passingham E.R., Colebatch G.J., Friston J.K., Nixon D.P., R S Frackowiak S.R.: Cortical areas and the selection of movement: a study with positron emission tomography. Experimental Brain Research. 1991;84(2):393-402. DOI: 10.1007/BF00231461.

[20] Nahab B.F., Hallet M.: Current Role of fMRI in diagnosis of movement disorders. Neuroimaging Clinics of North America. 2010; 20(1): 103-110. DOI:10.1016/j.nic.2009.08.001.

[21] Onozuka M., Fujita M., Watanabe K., Hirano Y., Niwa, M., Nishiyama K., Saito S. : Mapping brain region activity during chewing: a functional magnetic resonance imaging study. Journal of Dental Research. 2002; 81(11): 743-746. DOI.org/10.1177/0810743

[22] Kristeva R., Cheyne D., W.Lang W., Lindinger G., Deecke L.: Movement-related potentials accompanying unilateral and bilateral finger movements with different inertial loads. Electroencephalography and Clinical Neurophysiology.1990; 75(5): 410-418. DOI.org/10.1016/0013-4694(90)90086-Y

[23] Lund P.J. : Mastication and its Control by the Brain Stem. Critical Reviews in Oral Biology and Medicine. 1991; 2(1):33-64. DOI.org/10.1177/10454411910020010401

[24] Morimoto T., Hamada T., Kawamura Y. : Alteration in directional specificity of interdental dimension discrimination with the degree of mouth opening. Journal of Oral Rehabilitation.1983; 10(4): 335-342. DOI .org/10.1111/j.1365-2842.1983.tb00128.x

[25] Marr D.:A theory of cerebellar cortex. Journal of Physiology. 202(2): 437-470.1. DOI: 10.1113/jphysiol.1969.sp008820

[26] Yamazaki T., Lennon W.: Revisiting a theory of cerebellar cortex. Neuroscience Research. 2019; 148: 1-8. DOI.org/10.1016/j.neures.2019.03.001

[27] Ikeda A., Shibasaki H., Nagamine T., Terada K., Kaji R., Fukuyama H., Kimura J.: Dissociation between contingent negative variation and Bereitschaftspotential in a patient with cerebellar efferent lesion. Electroencephalogrphy Clinical Neurophysiology. 1994; 90 (5):359-364. DOI.org/10.1016/0013-4694(94)90051-5

[28] Sasaki K., Gemba H. , Hashimoto S.: Premovement slow cortical potentials on selfpaced hand movements and thalamocortical and corticocortical responses in the monkey. Experimental Neurology. 1981; 72(1): 41-50. DOI. org/10.1016/0014-4886(81)90125-4

[29] Naito E., EhrssonH.H.: Kinesthetic illusion of wrist movement activates motor-related areas. 2002; Neuroreport 12(17):3805-9. DOI: 10.1097/00001756-200112040-00041

Section 3

Aging and Dementia

Chapter 5

Mitochondria in the Cerebral and Cerebellar Cortex in Alzheimer's Disease, Target for a Therapeutic Approach

Stavros J. Baloyannis

Abstract

Alzheimer's disease remains the main cause of dementia in advanced age worldwide. Among the etiopathological background of the disease mitochondrial alterations may play a crucial role, given that they are closely related to metabolic and energy deficiency in neurons, glia, and endothelial cells in Alzheimer's disease and other neurodegenerative disorders. In a series of morphological and morphometric studies of mitochondria in the cerebrum and the cerebellar cortex in Alzheimer's disease, by electron microscopy, we described marked morphological and morphometric alterations. The most frequent ultrastructural alterations of the mitochondria consist of disruption of the cristae, accumulation of osmiophilic material, and marked changes of shape and size in comparison with the normal controls. Mitochondrial alterations were particularly prominent in dendritic profiles and dendritic spines. The ultrastructural study of a substantial number of neurons in the cerebellum revealed that mitochondrial alterations do not coexist, as a rule, with the typical Alzheimer's pathology, such as cytoskeletal alterations, amyloid deposits, and tau pathology, though they are frequently observed coexisting with alterations of the cisternae of the Golgi apparatus. Therapeutical regimes targeting mitochondria may be beneficial in early cases of Alzheimer's disease.

Keywords: Alzheimer's disease, Mitochondria, Electron microscopy, Oxidative stress Treatment, cerebrum, cerebellum

1. Introduction

Alzheimer's disease is the main causative factor of presenile and senile dementia [1] involving a large number of potential pathogenetic mechanisms, which for years was extinguishing the mental capacities, affecting seriously the cognition of the patients and leading to a tragic epilogue of the life with many social, economic and humanitarian consequences.

The phenomenology of familial or sporadic Alzheimer's disease is the final act of a drama, which gradually was causing selective and progressive neuronal loss [2], extensive synaptic alterations [3, 4], progressive neurofibrillary degeneration [5] resulting in intracellular accumulation of hyperphosphorylated tau protein [6], in the form of neurofibrillary tangles, with a parallel accumulation of extracellular deposits of Aβ peptide forming neuritic plaques with an obvious microglial involvement [7]. The accumulation of the Aβ peptide as the main causative factor in Alzheimer's disease has been the core of the amyloid cascade hypothesis, which gained a considerable reputation, attempting to interpret all the pathological phenomena in the stream of the morphological and functional disintegration in Alzheimer's disease [8].

However, a substantial body of evidence underlines the increasing differentiation from the amyloid hypothesis [9] and emphasizes the crucial role that mitochondrial alterations and dysfunction may play in the pathogenesis of Alzheimer's disease and other neurodegenerative disorders [10–14].

Mitochondria are double membraned organelles, which are the cardinal energy suppliers of the eukaryotic cells by generating ATP, via oxidative phosphorylation. Mitochondria have their circular, double-stranded DNA (mtDNA), encoding thirteen proteins essential for oxidative phosphorylation [15], which is continuously processed by five protein complexes of the respiratory chain (complexes I-V).

Mitochondrial DNA plays reasonably a crucial role in the homeostatic mechanisms of the cell, by providing the essential energetic background for most of the cellular procedures. Moreover, mitochondrial DNA is also involved in a significant number of functional pathways, concerning cellular signaling by generating reactive oxygen species (ROS) synthesis of neurotransmitters at the presynaptic terminals. It is reasonable, that based on their multidimensional activity, mitochondria would be versatile structures, continuously renewed by fusion and fission [16] and frail to degradation thru mitophagy [17, 18].

Mitochondrial morphology is mainly modulated by the neurofilaments and microtubules, given that mitochondria are mostly transported along the microtubules [19], expressing at the same time an immediate adaptation to energetic needs and immune responses of the cells [20]. The fact that mitochondria have an antiviral signaling protein (MAVS), in connection to the outer membrane, emphasizes their importance in activating immune reactions [21] and participating in antiviral responses [22].

In Alzheimer's disease, the mitochondrial alterations, are responsible for the reduced energy production, oxidative stress, and the inflammatory reactions [23], which are among the early phenomena of the disease [24, 25], in the broad spectrum of the functional and morphological alterations [26, 27], which occur affecting progressively the neuronal and synaptic integrity.

Mitochondrial alterations resulting in substantial oxidative stress have been described in a considerable number of neurodegenerative diseases [28–30], a fact which emphasizes the importance of mitochondria morphological and functional integrity in the normal life and long survival of neurons and glial cells. Oxidative stress triggers also the initiation of a real cascade of pathological phenomena, including the modulation of innate immunity, which provokes a further mitochondrial dysfunction, given that mitochondria and mtDNA are very sensitive to oxidative stress [31, 32]. Besides, the association of oxidative stress with the increased accumulation of calcium ions [33], would also be considered among the principal causes of apoptosis [34].

Oxidative stress in Alzheimer's disease is mostly related to amyloid β (Aβ) accumulation in the neocortex [35, 36], which is a phenomenon playing a crucial role in the pathogenetic process of Alzheimer's disease [37]. The mitochondrial dysfunction has culminated with the existent beta-Amyloid toxicity in connection with the decreased rate of glycolysis [38] and the inhibition of the mitochondrial cytochrome c oxidase by a dimeric conformer of Aβ42 [39, 40]. Besides, the increasingly synthesized reactive oxygen species (ROS), aggravate the mitochondrial dysfunction, increase the mitochondrial Ca load [41], initiating mitophagy eventually [42].

However, mitochondrial ROS in low levels may play a positive role acting as second messengers and controlling several physiological processes [43]. Also, ROS are considered as being responsible for NLRP3 inflammasome activation [44, 45].

In excessive ROS synthesis the endogenous antioxidant defense system, such as superoxide dismutase, glutathione peroxidase, superoxide reductase, catalase, are unable to counteract the ROS's vulnerability. It is also significant, that cytosolic mtDNA may activate the NLRP3 inflammasome increasing, even more, the inflammatory reactions [46].

Also, the hyperphosphorylated tau protein interacts with the voltage-dependent anion channel 1 (VDAC1) protein, affecting mitochondrial pores and deteriorating mitochondrial activity [47]. In a parallel way, caspase-cleaved tau impairs mitochondrial dynamics in Alzheimer's disease [48]. Therefore, it seems that the convergence of amyloid and tau pathology on mitochondria impair synergistically the mitochondrial function and exacerbate oxidative stress [49].

From the morphological point of view, the shape and the size of mitochondria are highly variable depending upon the fusion and fission processes, which are regulated by mitofusins (Mfn-1 and Mfn- 2) and optic atrophy protein- 1 (OPA- 1) [50]. From the morphometric point of view, the number of the mitochondria varies in the soma and neuronal processes, according to the energy state of the cell, given that they are transported and accumulated to regions where energy demands and ATP consumption are particularly high [51].

In Alzheimer's disease, morphological alterations of the mitochondria have been described [11, 52, 53] even in the early cases of the disease [11] coinciding with dendritic and synaptic pathology [54]. Some evidence suggests that the interaction of mitochondrial fission protein DRP- 1 with the Aβ peptide and the hyperphosphorylated tau protein results in mitochondrial fragmentation increasing therefore the mitochondrial damage [55].

Mitochondrial trafficking in Alzheimer's disease plays also an important role in abnormal mitochondrial positioning and accumulation. Mitochondrial motility is controlled normally by kinesin and dynein which are powered by ATP hydrolysis, whereas the immobilization of mitochondria in places of high energy consumption is controlled by syntaphilin [56]. In Alzheimer's disease, the hyperphosphorylated tau protein at AT8 sites (ROS), as by-products of the respiration chain, aggravates the mitochondrial dysfunction, increases the mitochondrial Ca load and the neuronal oxidative stress [41], initiating mitophagy eventually [42].

However, mitochondrial ROS in low levels may play a positive role acting as second messengers and controlling several physiological processes [43]. I addition, ROS are considered as being responsible for NLRP3 inflammasome activation [44, 45].

On the contrary, in excessive ROS synthesis the endogenous antioxidant defense system, is unable to counteract the ROS's vulnerability. It is also significant the fact, that cytosolic mtDNA may activate the NLRP3 inflammasome increasing, even more, the inflammatory reactions [46].

Besides the hyperphosphorylated tau protein interacts with the voltage-dependent anion channel 1 (VDAC1) protein, affecting mitochondrial pores and deteriorating mitochondrial function [47]. In a parallel way, caspase-cleaved tau impairs mitochondrial dynamics in Alzheimer's disease [48]. Therefore, it seems that the convergence of amyloid and tau pathology on mitochondria impair synergistically the mitochondrial activity and exacerbate oxidative stress [49].

From the morphological point of view, the shape and the size of mitochondria are highly variable depending upon the fusion and fission processes, which are regulated by mitofusins (Mfn-1 and Mfn-2) and optic atrophy protein-1 (OPA-1). [50]. From the morphometric point of view, the number of the mitochondria varies in the soma and neuronal processes, according to the energy state of the cell, given that they are transported and accumulated to regions where energy demands and ATP consumption are particularly high [51].

In Alzheimer's disease, morphological alterations of the mitochondria have been described [11, 52, 53] even in the early case of the disease [11] coinciding with dendritic and synaptic pathology [54]. Some evidence suggests that the interaction of mitochondrial fission protein DRP- 1 with the Aβ peptide and the hyperphosphorylated tau protein results in mitochondrial fragmentation increasing therefore the mitochondrial damage [55].

Mitochondrial trafficking in Alzheimer's disease plays also an important role in abnormal mitochondrial positioning and accumulation. Mitochondrial motility is controlled normally by kinesin and dynein which are powered by ATP hydrolysis, whereas the immobilization of mitochondria in places of high energy consumption is controlled by syntaphilin [56].

In Alzheimer's disease, the hyperphosphorylated tau protein at AT8 sites [57] may impede the mitochondrial transport via microtubules, leading to improper distribution of mitochondria in giant spines and abnormal synapses [58, 59]. In parallel, oligomers of Ab peptide impair mostly the anterograde movement of mitochondria [60], increasing the number of stationary mitochondria in the neuronal soma.

The observation that mitochondrial abnormalities occur as an early phenomenon in Alzheimer's disease [11], supports the hypothesis that mitochondrial degeneration plays a primal role in Alzheimer's disease pathogenetic procedure, inducing a chain of pathological alterations involving tau and amyloid pathology. Morphological alteration of mitochondria, as well as abnormal interconnections of mitochondria with neurofilaments and microtubules, have been described at the level of electron microscopy in dendrites, axons, and synaptic components, a fact which emphasizes the close association of mitochondrial pathology with the broad pattern of the morphological changes in Alzheimer's disease [61, 62].

In this study, which is an extensive observation on electron microscopy, we attempted to describe the morphological alterations of mitochondria in early cases of Alzheimer's disease in neurons from various areas of the cerebral and cerebellar cortex, proposing also therapeutic approaches in the initial stages of Alzheimer's disease, based on the existing mitochondrial pathology.

2. Material and methods

2.1 Material

For describing the morphological alterations of neuron's organelles in early cases of Alzheimer's disease by electron microscopy we focused our observation mostly on the mitochondria in twenty-two cases, fourteen men and eight women, aged 52–87 years, who fulfilled all the clinical, neuropsychological [63] and laboratory diagnostic criteria of Alzheimer's disease.

The brains were derived from patients, who died accidentally 24 to 46 months following the clinical diagnosis of Alzheimer's disease. Additional 15 brains, macroscopically intact derived from apparently healthy persons of parallel age with the patients, were used as normal controls.

Multiple samples from many areas of the brain, namely from the prefrontal area of the frontal lobe, the frontal pole, the acoustic cortex, the visual cortex, the parietal lobe, the insula, the vermis of the cerebellum, and the cerebellar hemispheres were taken in a room temperature of 4^0 C., 4 to 5 hours after death. Samples also from the hippocampus, the hypothalamus, the mammillary bodies, the locus coeruleus, the red nucleus, the globus pallidus were excised under the same conditions, processed for electron microscopy, and the findings were described and in previous reports.

2.2 Method

All the specimens were immediately immersed in Sotelo [64] fixing solution, for three hours, then post-fixed in osmium tetroxide for 30 min. and dehydrated in graded alcohol solutions and propylene oxide. Thin sections were cut in a Reichert ultratome, contrasted, with uranyl acetate and lead citrate, and studied in electron microscopes Elmiscope 1 and Zeiss 9As. All the methodological and technical details of the preparation of the specimens for electron microscopy have been described extensively in our previous reports [65, 66].

Following the morphological description of the mitochondria, we proceeded also to morphometric estimations on micrographs of a standard magnification of 56.000X.

The methodology of the morphometric estimation of the mitochondria and the statistical analysis of the data have been extensively described in our previous reports [11, 65].

3. Results

The mitochondria in cases of Alzheimer's disease demonstrate a wide variation of size and shape in comparison with the mitochondria of normal control brains (**Figure 1**). We noticed that numerous mitochondria were small round or elongated, particularly those which were inside the dendritic profiles or the synaptic terminals (**Figure 2**).

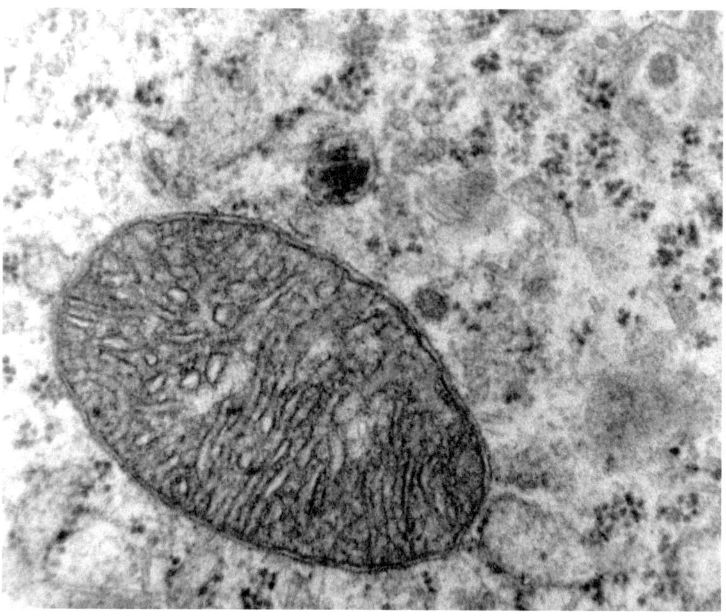

Figure 1.
Mitochondrion of a Purkinje cell of the cerebellum of 75 years old man unremarkable neurologically. (Mag. 72,000 X).

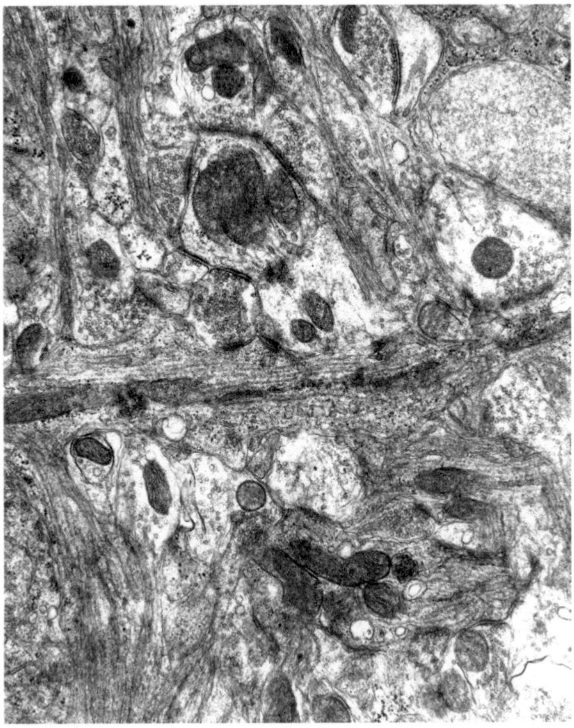

Figure 2.
Mitochondrial alterations in synaptic terminals and dendritic profiles in the molecular layer of the cerebellum of a male patient aged 75 years, suffered from Alzheimer's disease. The disruption of the mitochondrial cristae is obvious. Electron micrograph (Mag.68,000 X).

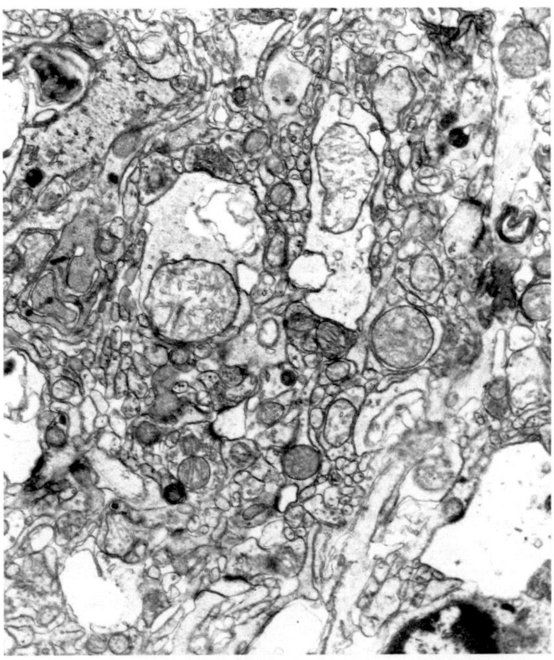

Figure 3.
In the majority of the synaptic profiles the mitochondria are elongated demonstrating an impressive polymorphism, concerning the arrangement of the cristae. Mitochondria in a synaptic profile in the molecular layer of the cerebellum of a male patient aged 68, who suffered from Alzheimer's disease. Electron micrograph (Mag. 68,000X).

DOI: http://dx.doi.org/10.5772/intechopen.98436

A substantial number of mitochondria show disruption of the cristae, though others include osmiophilic material [10, 11, 13]. In the majority of the synaptic profiles, the mitochondria showed an impressive polymorphism, concerning the pattern and the arrangement of the cristae (**Figure 3**). That polymorphism was particularly obvious in dendritic profiles in acoustic and visual cortices, where morphological alterations of the mitochondria coexisted frequently with the fragmentation of the Golgi apparatus (**Figure 4**).

The ultrastructural study of the cerebellar cortex, in the vermis and the hemispheres, revealed impressive mitochondrial polymorphism in the soma of the neurons, the dendritic profiles (**Figure 5**), as well as in the axons and the synaptic terminals (**Figure 6**) [66, 67].

Besides, the electron microscopy study revealed that morphological alterations are frequently seen in neurons of the prefrontal cortex, which included mostly small round mitochondria, with an abnormal arrangement of the cristae (**Figure 7**). Abnormal polymorphic mitochondria in association with the fragmentation of the Golgi apparatus were also observed in the Purkinje cells of the cerebellar cortex in the vermis and the hemispheres [68], in the stellate cells of the molecular layer of the cerebellar cortex (**Figure 8**), as well as in a substantial number of neurons of the prefrontal cortex [69].

By the morphological analysis of the mitochondria in the cortex of the brain hemispheres in Alzheimer's disease, it was realized that mitochondrial pathology was associated as a rule with dendritic and spinal pathology [70].

From the morphometric point of view, the ellipsoid mitochondria in normal controls appear to have an average diameter of 650 ± 250 nm and a mean axial ratio of 1.9 ± 0.2. The round or global mitochondria in normal controls appeared to have a mean mitochondrial radius of 350 nm.

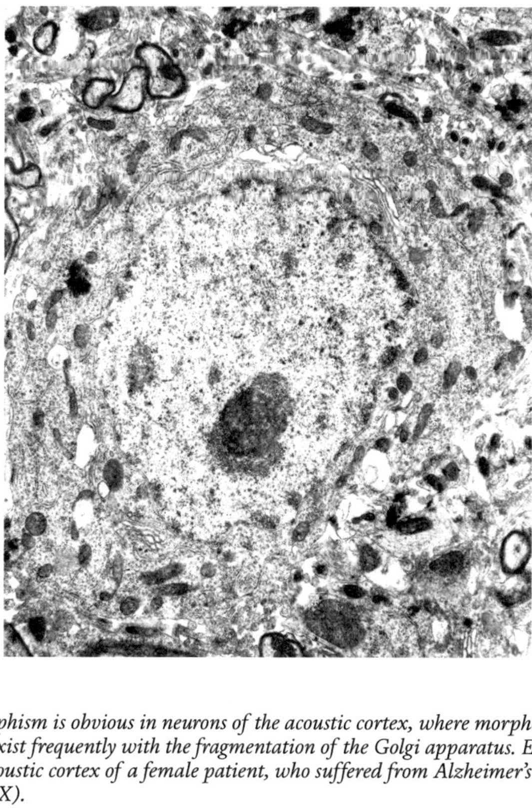

Figure 4.
Mitochondrial polymorphism is obvious in neurons of the acoustic cortex, where morphological alterations of the mitochondria coexist frequently with the fragmentation of the Golgi apparatus. Electron micrograph of a neuron from the acoustic cortex of a female patient, who suffered from Alzheimer's disease at the age of 73 years, (Mag. 28, 000 X).

In the brains of patients who suffered from Alzheimer's disease, the ellipsoid mitochondria of the neurons appeared to have an average diameter of 510 ± 250 nm and a mean axial ratio of 1.7 ± 0.2. The round mitochondria have had a mean radius of 280 nm. Also, the round mitochondria appeared to have a mean radius of 350 nm.

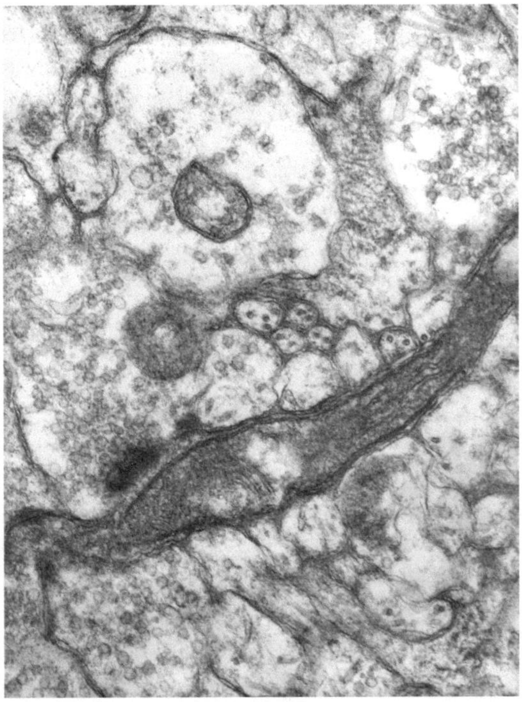

Figure 5.
Very elongated mitochondrion in a dendritic profile in the mocelular layer of the vermis of a male patient aged 63 years, who suffered from Alzheimer's disease in the early stages. Electron micrograph (Mag. 128,000 X).

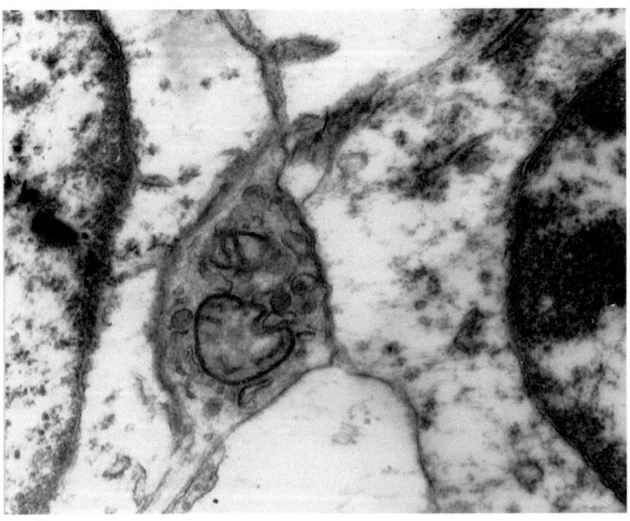

Figure 6.
Fragmentation of the cristae and impressive polymorphism of mitochondria in dendritic spines in the cortex of the cerebellar hemispheres of a male patient who suffered from Alzheimer's disease at the age of 75 years. Electron micrograph. Mag.135.000X).

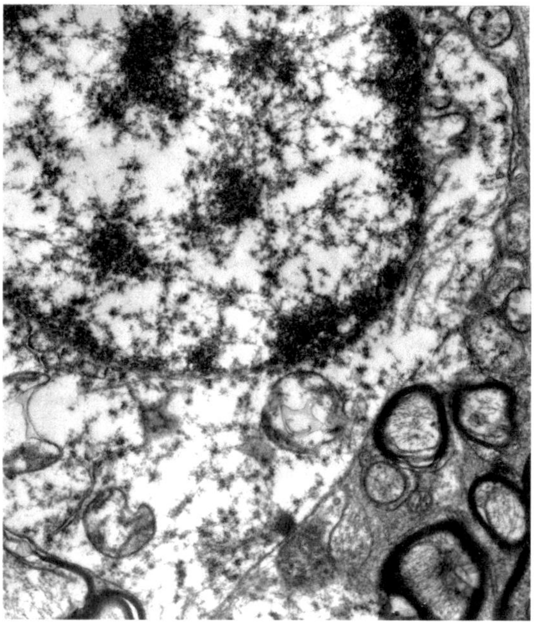

Figure 7.
Small mitochondria with abnormal arrangement of the cristae in a neuron from the prefrontal cortex of a male patient aged 75 years, who suffered from Alzheimer's disease in the early stages. Electron microgram (Mag. 28,000 X).

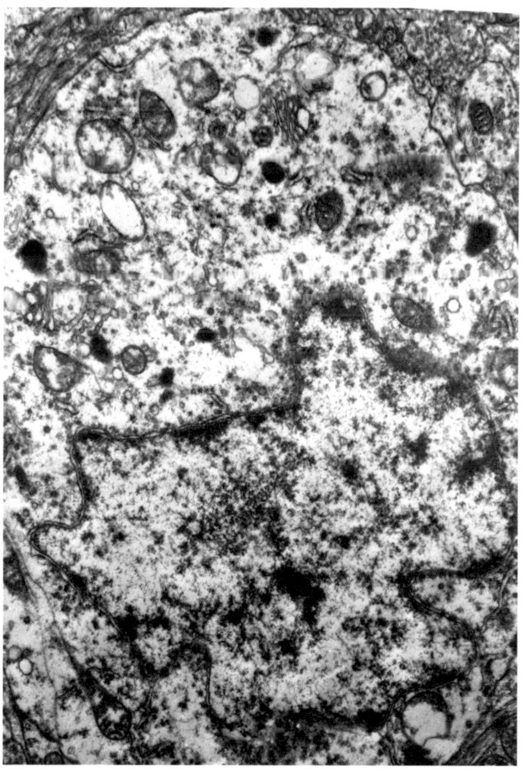

Figure 8.
Morphological alterations of mitochondria, coexist frequently with fragmentation of Golgi apparatus. Small compact mitochondria and fragmented cisternae of Golgi apparatus in the soma of an interneuron (stellate cell) of the molecular layer of the cerebellum of a male patient aged 63 years, who suffered from Alzheimer's disease in the early stages. Electron micrograph (Mag. 28,000 X).

4. Discussion

The morphological alteration of the mitochondria, which was extensively observed in the cortex of the brain hemispheres, the vermis and the hemispheres of the cerebellum pleads in favor of a generalized mitochondrial dysfunction in Alzheimer's disease, which would be associated with wide neuronal loss, impaired axoplasmic flow, dendritic pathology, and marked synaptic and spinal alterations, which would be seriously affecting the mental faculties of the patients [71].

The defective mitochondria in Alzheimer's neurons may not supply adequate levels of Adenosine Triphosphate (ATP), which is very important at the synaptic level for normal neural communication. It is expectable that the low levels of cellular ATP at nerve terminals may lead to extensive loss of synapses or cause defective function in the majority of them [72]. Besides, oxidative stress decreases the rate of choline recycling at the synapses, leading to Ach deficiency [73].

Many morphological alterations of AD could be linked to mitochondria changes since blockage of mitochondrial energy production shifts amyloid-protein precursor metabolism to the production of more amyloidogenic forms of amyloid [74]. Thus it induces the production of A68 antigen [75, 76], and activates the mitogen-activated protein kinase pathway [77–79].

Also, inadequate energy production impairs the mitochondrial motility in the soma, the axons, and the dendritic branches of neurons, resulting in trafficking jams aggravating even farther the mitochondrial function [80].

Accumulation also of transmembrane-arrested AβPP may block protein translocation, affecting, even more, the mitochondrial function. In a parallel way, the accumulated Aβ peptide in the mitochondrial membrane may be transported from the cytosol via mitochondrial translocases, which are located either in the outer or the inner mitochondrial membranes. Moreover, the Aβ peptide interacts with an Aβ-binding dehydrogenase (ABAD) in the mitochondria of patients suffering from Alzheimer's disease as well as in transgenic mice, suggesting that ABAD is closely related, to mitochondrial toxicity [81]. Overexpression of ABAD can increase oxidative stress, accelerating, therefore, neuronal death. However, ABAD may play an important role in the oxidation of alcohols, facilitating the reduction of aldehydes and ketones, and decreasing subsequently the metabolic stress [82].

The Aβ-peptide may interact with cyclophilin D (CypD), a component of the mitochondrial transition pore, inducing cytotoxicity [83]. Moreover, morphological alterations of mitochondria in AD may be related to the increased mitophagy, which is proved by the accumulation of mitochondrial autophagic elements in neurons of AD patients [84]. In AD the PINK1-Parkin-dependent mitophagy pathway may also be involved in mitochondrial pathology [85]. The prompt clearance of damaged mitochondria may result in increasing the density of normal mitochondria in dendrites and synaptic terminals, which is a fact ameliorating the synaptic function [86].

5. Suggestions on the treatment of Alzheimer's disease on the basis of mitochondrial pathology

Concerning the treatment of Alzheimer's disease, we would underline that the preclinical stage is frequently overlooked, because it might be characterized as mild cognitive impairment, with considerable consequences on the course and the treatment of the disease.

Following the clinical manifestation of the disease and the diagnostic documentation, many therapeutic regimes have been applied without any substantial beneficial effect. In the decade 2002–2012 more than 240 drugs, mostly cholinesterase inhibitors, and NMDA receptor antagonists, have been tried for the treatment or even the amelioration of the quality of life in patients suffered from AD [87, 88], without any obvious effectiveness. Besides, any strategy attempting to reduce the amyloid aggregations in the brain, despite the numerous trials, was not fruitful [89].

Based on mitochondrial pathology, in the limits of the broad pathogenetic spectrum of Alzheimer's disease, the mitochondria may be considered as the potential therapeutic targets, which might inhibit the stream of the neuropathological alterations and impede the clinical deterioration of the patients.

Strategists protecting the mitochondria in Alzheimer's disease would include the administration of efficient antioxidant factors, which might counteract the oxidative stress, and decrease ROS production [90].

Natural antioxidants that could penetrate the blood–brain barrier may be effective in the initial stage of Alzheimer's disease. The administration of Vitamins C, E, beta carotene, glutathione, Coenzyme Q10, epigallocatechin gallate, curcumin, lipoic acid, *Ginkgo biloba*, resveratrol, pramipexole, N-acetylcysteine, latrepirdine, idebenone, ubiquinone may reduce the production of ROS, suppressing the oxidative stress.

The tetracyclin Minocycline prevents also oxidative stress and controls the release of cytochrome c from mitochondria, inhibiting the activation of caspase-3 and the subsequent apoptosis [91], being therefore quite effective in the treatment of AD [92].

In a parallel way, the adaptation of the Cretic or Mediterranean diet, combined with frequent proper physical exercise, in the early stages of the disease, may stabilize the mental capacities of the patients for a non-limited period and postpone the tragic epilogue of the disease. Besides, prolonged administration of pyruvate may improve the working memory in the preclinical stage AD [93].

We have realized, by a detailed clinical and neuropsychological evaluation of a considerable number of patients, who suffered from Alzheimer's disease in the initial stages, that the quotidian administration of Riboflavin (Vit.B2) in a dose of 100–200 mg per day would play a positive role in inhibiting the course of the disease and stabilizing the mental faculties of the patients (Baloyannis, unpublished data). It is known that riboflavin serves as a flavoprotein precursor in the synthesis of flavin adenine dinucleotide (FAD) and flavin mononucleotide (FMN) [94, 95], which are electrochemically-active factors involved in regulatory pathways of mitochondria. Riboflavin serves also as a cofactor in fatty acid β-oxidation. Riboflavin deficiency may be involved in the pathogenetic mechanism of several neurodegenerative disorders [96, 97]. Thus, riboflavin supplementation may be enriching the therapeutical regime in some non-uncommon neurological conditions [98, 99] including Alzheimer's disease [100].

At the experimental level, mitochondrial-targeted molecules, such as MitoQ and Szeto- Schiller (SS) peptides have been used for increasing the concentration of antioxidants into mitochondria [101]. MitoQ exerts direct antioxidant action by scavenging superoxide, peroxyl, and peroxynitrite ROS. It seems also to contribute effectively, enhancing the mitochondrial biogenesis in a transgenic mouse model of Alzheimer's disease [102, 103].

The fact that the interaction of ABAD with the Aβ peptide may increase the toxic effects upon the mitochondria, suggests that ABAD inhibitors such as AG18051 [104] and RM-532-46, might be applied as therapeutic factors in the treatment of patients who suffer from AD.

Also, a decrease of the CypD in AD relieves the toxic effect that Aβ imposes on the mitochondria and may ameliorate the mental condition of the patients [83] improving the synaptic function by the restoration of mitochondrial activity [105]. Also, Oligomycin-sensitivity conferring protein (OSCP) is a crucial subunit of mitochondrial F1Fo ATP synthase, essential for its structural stability [106]. In Alzheimer's disease, deregulation of mitochondrial F1FO-ATP synthase was described [107]. It is reasonable to be hypothesized that OSCP would be a positive factor in the treatment of early cases of Alzheimer's disease.

The administration of factors that may regulate mitophagy, and cardiolipin-induced mitophagy [108] such as NAD+ precursors, actinonin (AC), spermidine, urolithin A (UA), rapamycin, and doxycycline may also control the mitochondrial unfolded protein response (UPRmt), reducing the Aβ accumulation and proteotoxicity [109].

The administration of metformin can increase also the resistance of the mitochondria to oxidative stress [110]. The supplementation with nicotinamide may be beneficial in the early stages of Alzheimer's disease improving the cognition of the patients [111].

The administration of mitochondrial uncoupling factors such as 2, 4-dinitrophenol (DNP) may be effective too, protecting mitochondria and stabilizing neuronal function in animal models of AD [112, 113]. Besides, the administration of galanthamine hydrochloride may control autophagy [114], as was noticed by the decrease of autophagosome formation.

Recently it was found that vacuole membrane protein 1 (VMP1), which is located in the endoplasmic reticulum (ER), may play a crucial role in mediating autophagy [115] and controlling mitochondrial morphology, given that numerous mitochondria are damaged upon VMP1 deficiency [116]. It must be underlined that autophagy is an important mechanism for maintaining cell homeostasis by liberating the cell from the accumulation of misfolded proteins and other undesired elements [117].

Erythropoietin (EPO) [118], is a cytokine essential for erythroid development and maturation, playing a beneficial role in progressive degenerative diseases [119] exercising among others a protective effect on mitochondrial morphology, facilitating the activity of cellular bioenergetics [120].

Furthermore, the relation between mitochondria and endoplasmic reticulum (ER) referred to as the MAMs [121, 122], which are enriched in presenilin proteins [123], may play an important role in the pathogenetic cascade of Alzheimer's disease, given that they are involved in the production of intracellular Aβ [124], and play also a substantial role in cellular Calcium homeostasis [125, 126] and lipid transport between the endoplasmic reticulum and the mitochondria [127]. A deep understanding of MAMs involvement in the pathogenesis of AD may provide new ways of therapeutic approach in the early stages of AD.

The protection of mitochondria by the Szeto-Schiller (SS) peptides, particularly the SS31 tetra-peptide [128, 129], which is targeted to the inner mitochondrial membrane, revealed that it may have protective effects against mitochondrial and synaptic toxicities in APP transgenic mice [130], reducing also mitochondrial fragmentation and increasing mitochondrial transport in AD neurons [131].

Mitochondrial biogenesis is decreased in AD [132] due to the decrease of the rate of mitochondrial division, a fact that aggravates substantially the mitochondrial dysfunction [133]. Supporting mitochondrial biogenesis may contribute to the therapeutic confrontation of AD [134]. Therefore, the administration of Nicotinamide riboside [135, 136], as well as of pioglitazone or rosiglitazone may improve the mental condition of the patients [137, 138].

In experimental models, the administration of melatonin contributed considerably to improving the biogenesis of mitochondria [139]. Besides, melatonin inhibits amyloidogenesis and promotes the non-amyloidogenic pathway [140], restoring also the equilibrium of the $Ca2+$ [141]. Recent neuropathological findings revealed that melanin-concentrating hormone (MCH) neurons in the lateral hypothalamic area of patients who suffer from AD undergo degeneration, a fact that may interpret the frequent sleep and metabolic disorders of the patients [142].

Also, it was observed, that the Zinc ion ($Zn2+$) supplementation in experimental models, contributed to ameliorating the mitochondrial function by the restoration of BDNF levels and improving cognition, although it might have negative effects on some cells lines [143].

A substantial body of evidence suggests that among the causative factors in the multifactorial labyrinth of AD, the hemodynamic disturbances resulting in chronic hypoperfusion of the brain [144–147] play also a crucial role in attenuating the mitochondrial dysfunction [148] and aggravating subsequently the mental condition of the patients [149]. The improvement of the blood supply of the brain should be among the principal therapeutic strategists in AD.

Mitochondrial alterations may be estimated as potential biomarkers, which would provide a prognostic response to treatment [150]. Mitochondria may be considered as a significant strategic point for therapeutical interventions in early cases of Alzheimer's disease.

6. Conclusions

Mitochondrial alterations may play an important role in the pathogenesis of Alzheimer's disease. In early cases of Alzheimer's disease marked morphological and morphometric alterations have been described by electron microscopy in various areas of the brain and the cerebellum.

The morphological alterations of the mitochondria seem to be independent of Alzheimer's pathology, given that they are observed in areas without or with minimal neuritic plaques or tau pathology.

Mitochondria may be considered as a significant strategic point for therapeutical interventions in early cases of Alzheimer's disease.

Conflict of interest

The author declares no conflict of interest.

Author details

Stavros J. Baloyannis
Department of Neurology, Aristotelian University of Thessaloniki, Thessaloniki, Greece

*Address all correspondence to: sibh844@otenet.gr

References

[1] Prince M, Bryce R, Albanese E, Wimo A, Ribeiro W, Ferri CP. The global prevalence of dementia: A systematic review and meta-analysis. Alzheimers Dement. 2013; 9: 63– 75 e2.

[2] Terry R. The pathogenesis of Alzheimer disease: An alternative to the amyloid hypothesis. J. Neuropathology Exper. Neurology. 1996; 55: 1023-1025.

[3] Baloyannis S, Costa V, Arnaoutoglou A, Arnaoutoglou H. Synaptic alterations in the molecular layer of the cerebellum in Alzheimer's disease. Neuropath. Appl. Neurobiol. 1996; 22: 78-79.

[4] Baloyannis S, Costa V. Abnormal synapses in the granule cell layer of the cerebellum in Alzheimer's disease: A Golgi and Electron microscope study. European J. Neurol. 1998; 5: 26 27.

[5] Braak H, Braak E. Staging of Alzheimer's disease-related neurofibrillary changes. Neurobiol. Aging. 1995; 16: 271– 278.

[6] Morris M, Maeda S, Vossel K, Mucke L. The many faces of tau. Neuron, 2011; 70: 410– 426.

[7] Crass P, Kawai M, Siedlak S, Mulvihill P, Gambetti P, Loweryd D, Gonzalez De Whitt P, Greenberg B, Perry G. Neuronal and microglial involvement in β amyloid protein deposition in Alzheimer's disease. Am. J. Pathol. 1990; 37: 241 246.

[8] Hardy J A, Higgins GA. Alzheimer's disease: The amyloid cascade hypothesis. Science. 1992; 256: 184– 185.

[9] Drachman DA. The amyloid hypothesis, time to move on: Amyloid is the downstream result, not cause, of Alzheimer's disease. Alzheimers Dement. 2014; 10: 372– 380.

[10] Baloyannis S. Mitochondrial alterations in Alzheimer's Disease. Neurobiol. Aging. 1998; 19: S241.

[11] Baloyannis SJ. Mitochondrial alterations in Alzheimer's disease. J. Alzh. Dis. 2006; 9: 119– 126.

[12] Swerdlow RH, Burns JM, Khan SM. The Alzheimer's disease mitochondrial cascade hypothesis: Progress and perspectives. Biochim. Biophys. Acta. 2014; 1842: 1219-1231.

[13] Baloyannis S. Mitochondrial alterations in neurons of various areas of the brain in Parkinson's disease: an electron microscopy study. Parkinsonism and Related disorders. 2001; 7: S10.

[14] Tucker D, Lu Y, Zhang Q. From Mitochondrial Function to Neuroprotection-an Emerging Role for Methylene Blue. Mol. Neurobiol. 2018, 55,137-153.

[15] Cardoso S, Carvalho C, Correia S C, Seiça RM, Moreira PI. Alzheimer's disease: From mitochondrial perturbations to mitochondrial medicine. Brain Pathol. 2016; 26: 632– 647.

[16] Bertholet AM, Delerue T, Millet AM, Moulis MF, David C, Daloyau M, Belenguer P. Mitochondrial fusion/fission dynamics in neurodegeneration and neuronal plasticity. Neurobiol. Dis. 2016; 90: 3– 19.

[17] Rodolfo C, Campello S, Cecconi F. Mitophagy in neurodegenerative diseases. Neurochem. Internatl. 2017; 117: 156– 166.

[18] Barnhart EL. Mechanics of mitochondrial motility in neurons. Curr. Opin. Cell Biol. 2016; 38: 90– 99.

[19] Seo AY, Joseph AM, Dutta D, Hwang JC, Aris JP, Leeuwenburgh C. New insights into the role of mitochondria in aging: Mitochondrial dynamics and more. J. Cell Sci. 2010; 123: 2533– 2542.

[20] Sandhir R, Halder A, Sunkaria A. Mitochondria as a centrally positioned hub in the innate immune response. Biochim. Biophys. Acta Mol. Basis Dis. 2017; 1863: 1090-1097.

[21] Jacobs JL, Coyne CB. Mechanisms of MAVS regulation at the mitochondrial membrane. J. Mol. Biol. 2013; 425: 5009-5019.

[22] West AP, Shadel GS, Ghosh S. Mitochondria in innate immune responses. Nat. Rev. Immunol. 2011; 11: 389-402.

[23] Rimessi A, Previati M, Nigro F, Wieckowski MR, Pinton P. Mitochondrial reactive oxygen species and inflammation: Molecular mechanisms, diseases and promising therapies. Int. J. Biochem. Cell Biol. 2016; 81: 281-293.

[24] Wang X, Wang W, Li L, Perry G, Lee HG, Zhu X. Oxidative stress and mitochondrial dysfunction in Alzheimer's disease. Bioch. Biophys. Acta. 2014; 1842: 1240– 1247.

[25] Nunomura A, Perry G, Aliev G, Hirai K, Takeda A, Balraj EK, Jones PK, Ghanbari H, Wataya T, Shimohama S, Chiba S, Atwood CS, Petersen RB, Smith MA. Oxidative damage is the earliest event in Alzheimer disease. J. Neuropathol. Exp. Neurol. 2001;60:759-767.

[26] Guo C, Sun L, Chen X, Zhang D. Oxidative stress, mitochondrial damage and neurodegenerative diseases. Neural. Regen. Res. 2013; 8:2003-2014.

[27] Birnbaum JH, Wanner D, Gietl AF, Saake A, Kündig TM, Hock C, Nitsch RM, Tackenberg C. Oxidative stress and altered mitochondrial protein expression in the absence of amyloid-β and tau pathology in iPSC-derived neurons from sporadic Alzheimer's disease patients. Stem Cell Res. 2018; 27:121-130.

[28] Hirai K, Aliev G, Nunomura A, Fujioka H, Russell R, Atwood C, Johnson A, Kress Y, Vinters H, Tabaton M, Shimohama S, Cash A, Siedlak S, Harris P, Jones P, Petersen R, Perry G, Smith M. Mitochondrial Abnormalities in Alzheimer's Disease. J. Neurosci. 2001; 21: 3017-3023.

[29] Baloyannis S, Costa V, Michmizos D. Mitochondrial alterations in Alzheimer's Disease. Am. J. Alzheimers Dis. Other Demen. 2004; 19: 89-93.

[30] Mizuno Y, Ikebe S, Hattori N, Nakagawa-Hattori, Y, Mochizuki H, Tanaka M, Ozawa Y. Role of mitochondria in the etiology and pathogenesis of Parkinson's disease. Bioch. Biophys. Acta. 1995; 1271: 265-274.

[31] Margineantu D, Cox W, Sundell L, Sherwood S, Beechem J, Capaldi R. Cell cycle dependent morphology changes and associated mtDNA redistribution in mitochondria of human cell lines. Mitochondrion. 2002; 1: 425-435.

[32] Mathew A, Lindsley TA, Sheridan A, Bhoiwala DL, Hushmendy SF, Yager EJ, Ruggiero EA, Crawford DR. Degraded mitochondrial DNA is a newly identified subtype of the damage associated molecular pattern (DAMP) family and possible trigger of neurodegeneration. J. Alzheimers Dis. 2012; 30: 617-627.

[33] Khodorov B, Pinelis V, Vergun O, Storozhevykh T, Vinskaya N. Mitochondrial deenergization underlies neuronal calcium overload following a prolonged glutamate challenge. FEBS Lett. 1996; 397: 230-234.

[34] Duchen M. Contributions of mitochondria to animal physiology: from homeostatic sensor to calcium signaling and cell death. J. Physiol. (Lond). 1999; 516: 1-17.

[35] Morais Cardoso S, Swerdlow R, Oliveira C. Induction of cytochrome c-mediated apoptosis by amyloid beta 25-35 requires functional mitochondria. Brain Res. 2002; 931: 117-125.

[36] Moreira P, Santos M, Moreno A, Oliveira C. Amyloid beta-peptide promotes permeability transition pore in brain mitochondria. Biosci. Res. 2001; 21: 789-800.

[37] Pereira C, Santos M, Oliveira C. Involvement of oxidative stress on the impairment of energy metabolism induced by A beta peptides on PC12 cells: protection by antioxidants. Neurobiol. Dis. 1999; 6: 209-219.

[38] Arias C, Montiel T, Quiroz-Baez R, Massieu L. beta-Amyloid neurotoxicity is exacerbated during glycolysis inhibition and mitochondrial impairment in the rat hippocampus in vivo and in isolated nerve terminals: implications for Alzheimer's disease. Exp. Neurol. 2002; 176: 163-174.

[39] Crouch P, Blake, R.; Duce, J.; Ciccotosto, G.; Li, Q.; Barmham, K.; Curtain, C.; Cherny, R.; Cappai R, Dyrks T, Masters C, Trounce I. Copper-dependent inhibition of human cytochrome c oxidase by dimeric conformer of amyloid-1-42. J. Neurosci. 2005; 25: 672-679.

[40] Bosetti F, Brizzi F, Barogi S, Mancuso M, Siciliano G, Tendi E, Murri L, Rapoport SI, Solaini G. Cytochrome c oxidase and mitochondrial F1F0-ATPase (ATP synthase) activities in platelets and brain from patients with Alzheimer's disease. Neurobiol. Aging. 2002; 23: 371-376.

[41] Angelova PR, Abramov AY. Role of mitochondrial ROS in the brain: from physiology to neurodegeneration. FEBS Letters. 2018; 92: 692-702.

[42] Wang Y, Nartiss Y, Steipe B, McQuibban GA, Kim, P.K. ROS-induced mitochondrial depolarization initiates PARK2/PARKIN-dependent mitochondrial degradation by autophagy. Autophagy. 2012, 8, 1462-1476.

[43] Ristow M, Schmeisser K. Mitohormesis: Promoting Health and Lifespan by Increased Levels of Reactive Oxygen Species (ROS). Dose Response. 2014; 12: 288-341.

[44] Schroder K, Zhou R, Tschopp J. The NLRP3 inflammasome: A sensor for metabolic danger? Science. 2010; 327: 296-300.

[45] Shadel GS, Horvath TL. Mitochondrial ROS signaling in organismal homeostasis. Cell. 2015; 163: 560-569.

[46] Shimada K, Crother TR, Karlin J, Dagvadorj J, Chiba N, Chen S, Ramanujan VK, Wolf AJ, Vergnes L, Ojcius DM, Rentsendorj A, Vargas M, Guerrero C, Wang Y, Fitzgerald KA, Underhill DM, Town T, Arditi M. Oxidized mitochondrial DNA activates the NLRP3 inflammasome during apoptosis. Immunity. 2012;36:401-314.

[47] Manczak M, Reddy PH. Abnormal interaction of VDAC1 with amyloid beta and phosphorylated tau causes mitochondrial dysfunction in Alzheimer's disease. Human Molecular Genetics. 2012; 21: 5131– 5146.

[48] Perez MJ, Vergara- Pulgar K, Jara C, Cabezas- Opazo F, Quintanilla RA. Caspase- cleaved tau impairs mitochondrial dynamics in Alzheimer's disease. Mol. Neurobiol. 2018; 55: 1004-1018.

[49] Correia SC, Moreira PI, Perry G. Unraveling the Role of Mitochondria in Alzheimer's Disease In: Vascular Disease, Alzheimer's Disease, and Mild Cognitive Impairment. Edited by: David J. Libon, Melissa Lamar, Rodney A. Swenson, Kenneth M. Heilman, Oxford University Press. 2020. DOI: 10.1093/oso/9780190634230.003.0017.

[50] Ni HM,Williams JA, Ding WX. Mitochondrial dynamics and mitochondrial quality control. Redox Biol. 2015; 4; 6-13.

[51] Bereiter-Hahn J, Vöth M. Dynamic of mitochondria in living cells: shape changes, dislocations, fusion, and fission of mitochondria. Microsc. Res. Tech. 1994; 27: 198-219.

[52] Hirai K, Aliev G, Nunomura A, Fujioka H, Russell RL, Atwood C S, et al. Mitochondrial abnormalities in Alzheimer's disease. J. Neurosci. 2001; 21: 3017– 3023.

[53] Baloyannis SJ, Costa V, Michmizos D. Mitochondrial alterations in Alzheimer's disease. Am. J. Alzheimers Dis. Other Demen. 2004; 19: 89-93.

[54] Baloyannis SJ, Manolides SL, Manolides LS. Dendritic and spinal pathology in the acoustic cortex in Alzheimer's disease: morphological estimation in Golgi technique and electron microscopy. Acta Otolaryngol. 2011; 131: 610-612.

[55] Manczak M, Reddy PH. Abnormal interaction between the mitochondrial fission protein Drp1 and hyperphosphorylated tau in Alzheimer's disease neurons: Implications for mitochondrial dysfunction and neuronal damage. Hum. Mol. Genet. 2012; 21: 2538– 2547.

[56] Lin MY, Sheng ZH. Regulation of mitochondrial transport in neurons. Exper. Cell Res. 2015; 334: 35-44.

[57] Shahpasand K, Uemura I, Saito T, Asano T, Hata K, Shibata K, Hisanaga S. Regulation of mitochondrial transport and inter- microtubule spacing by tau phosphorylation at the sites hyperphosphorylated in Alzheimer's disease. J. Neurosci. 2012; 32:2430-2441.

[58] Baloyannis SJ. Dendritic pathology in Alzheimer's disease. J. Neurol. Sci. 2009; 283: 153-157.

[59] Baloyannis SJ, Mauroudis I, Manolides SL, Manolides LS. Synaptic alterations in the medial geniculate bodies and the inferior colliculi in Alzheimer's disease: a Golgi and electron microscope study. Acta Otolaryngol. 2009; 129:416-418.

[60] Quintanilla RA, Dolan PJ, Jin YN, Johnson GVW. Truncated tau and Abeta cooperatively impair mitochondria in primary neurons. Neurobiol. Aging. 2012; 33: 619 e25-35.

[61] Baloyannis SJ. Golgi apparatus and protein trafficking in Alzheimer's disease. J. Alzheimers Dis. 2014; 42 (Suppl 3):S153-162. doi: 10.3233/JAD-132660. PMID: 24946873.

[62] Baloyannis SJ, Mavroudis I, Baloyannis JS, Costa VG. Mammillary Bodies in Alzheimer's Disease: A Golgi and Electron Microscope Study. Am. J. Alzheimers Dis. Other Demen. 2016; 31: 247-256.

[63] McKhann G, Drachman D, Folstein M, Katzman R, Price D, Stadlan EM. Clinical diagnosis of Alzheimer's disease: report of the NINCDS-ADRDA work group under the auspices of department of health and human services task force on Alzheimer's disease. Neurology. 1984; 34: 939-944.

[64] Sotelo JR. Technical improvements in specimen preparation for electron microscopy. Exp. Cell Res. 1957; 13: 599-601.

[65] Baloyannis SJ. Recent progress of the Golgi technique and electron microscopy to examine dendritic pathology in Alzheimer's disease. Future Neurol. 2013; 8: 239-242.

[66] Baloyannis SJ, Mavroudis I, Mitilineos D, Baloyannis JS, Costa VG. The Hypothalamus in Alzheimer's disease: a Golgi and electron microscope study.Am. J. Alzheimers Dis. Other Demen. 2015; 30: 478-487.

[67] Baloyannis S, Manolidis S, Manolidis L. Synaptic alterations in the Vestibulocerebellar System in Alzheimer´s Disease- A Golgi and Electron Microscope Study. Acta Otolaryngol. (Stockh) 2000; 120: 247-250.

[68] Baloyannis S. The Golgi apparatus of Purkinje cells in Alzheimer's disease In Jurgen Bohl (Ed) Neuropathology- Back to the roots Shaker Vertag, Aachen, 2002, pp. 1-10.

[69] Baloyannis S, Manolidis S, Manolidis L. The acoustic cortex in Alzheimer's disease. Acta Otolaryngol. (Stockh). 1992; Suppl. 494: 1-13.

[70] Baloyannis SJ, Manolides SL, Manolides LS. Dendritic and spinal pathology in the acoustic cortex in Alzheimer's disease: morphological estimation in Golgi technique and electron microscopy. Acta Otolaryngol. 2011;131: 610-612.

[71] Mesulam M. Large-scale neurocognitive networks and distributed processing for attention, language and memory. Ann. Neurol. 1990; 28: 597-613.

[72] Baloyannis S, Theocharidis T, Manolidis L. Synaptic alterations in the acoustic cortex of the rat following insulin-induced hypoglycemia. Arch Otorhinolaryngol. 1987; 244: 36-43.

[73] Miar A, Alvarez V, Corao AI, Alonso B, Díaz M, Menéndez M, Martínez C, Calatayud M, Morís G, Coto E. Lack of association between protocadherin 11-X/Y (PCDH11X and PCDH11Y) polymorphisms and late onset Alzheimer's disease. Brain Res. 2011;1383: 252-256.

[74] Gabuzda D, Busciglio J, Chen L, Matsudaira P, Yankner B. Inhibition of energy metabolism alters the processing of amyloid precursor protein and induces a potentially amyloidogenic derivative. J. Biol. Chem. 1994; 269: 13623-13628.

[75] Blass J, Baker A, Ko L, Black, R. Induction of Alzheimer antigens by an uncoupler of oxidative phosphorylation. Arch. Neurol. 1990; 47: 864-869.

[76] Blass J, Fheu R, Gibson G. Inheritent abnormalities in energy metabolism in Alzheimer disease: Interaction with cerebrovascular compromise. Ann.N.Y. Acad. Sci. 2000; 903: 204-221.

[77] Luo Y, Bond J, Ingram V. Compromised mitochondrial function leads to increased cytosolic calcium and to activation of MAP kinases. Proc. Natl. Acad. Sci. USA. 1997; 94: 9705-9710.

[78] Perry G, Roder H, Nunomura A, Takeda A, Friedlich A, Zhu X, Raina A, Holbrook N, Siedlak S, Harris P, Smith M. Activation of neuronal extracellular receptor kinase (ERK) in Alzheimer disease links oxidative stress to abnormal phosphorylation. NeuroReport. 1999; 10: 2411-2415.

[79] Zhu X, Rottkamp C, Boux H, Takeda A, Perry G, Smith M. Activation of p38 pathway links tau phosphorylation, oxidative stress and cell cycle related events in Alzheimer disease. J. Neuropathol. Exp. Neurol. 2000; 59: 880-888.

[80] Correia SC, Perry G, Moreira PI. Mitochondrial traffic jams in Alzheimer's disease- pinpointing the roadblocks. Bioch. Biophys. Acta. 2016; 1862: 1909-1917.

[81] Lustbader J, Cirilli M, Lin C, Xu H, Takuma K, Wang N, Caspersen C, Chen X, Pollak S, Chaney M, Trinchese F, Liu S, Gunn-Moore F, Lue L, Walker D, Kuppusamy P, Zewier Z, Arancio O, Sten D, Yan S, Wu H. ABAD directly links Abeta to mitochondrial toxicity in Alzheimer's disease. Science. 2004; 304: 448-453.

[82] Benek O, Musílek K, Kuca K. Mitochondrial enzyme ABAD and its role in the development and treatment of Alzheimer's disease. Ceska Slov. Farm. Cas. Ceske Farm. Spolecnosti Slov. Farm. Spolecnosti, 2012; 61: 144-149.

[83] Du H, Guo L, Fang F, Chen D, Sosunov AA, McKhann GM, Yan Y, Wang C, Zhang H,Molkentin JD, Gunn-Moore FJ, Vonsattel JP, Arancio O, Chen JX, Yan SD. Cyclophilin D deficiency attenuates mitochondrial and neuronal perturbation and ameliorates learning and memory in Alzheimer's disease. Nat. Med. 2008; 14: 1097-1105.

[84] Moreira PI, Siedlak SL,Wang X, Santos MS, Oliveira CR, Tabaton M, Nunomura A, Szweda LI, Aliev G, Smith MA, Zhu X, Perry G. Increased autophagic degradation of mitochondria in Alzheimer disease. Autophagy. 2007; 3: 614-615.

[85] Martin-Maestro P, Gargini R, Perry G, Avila J, Garcia-Escudero V. PARK2 enhancement is able to compensate mitophagy alterations found in sporadic Alzheimer's disease. Hum. Mol. 2016; 25: 792-806.

[86] Fang EF, Hou Y, Palikaras K, Adriaanse BA, Kerr JS, Yang B, Lautrup S, Hasan-Olive MM, Caponio D, Dan X, Dan X, Rocktäschel P, Croteau DL, Akbari M, Greig NH, Fladby T, Nilsen H, Cader MZ, Mattson MP, Tavernarakis N, Bohr VA. Mitophagy inhibits amyloid-beta and tau pathology and reverses cognitive deficits in models of Alzheimer's disease. Nat. Neurosci. 2019; 22: 401-412.

[87] Cummings JL, Morstorf T, Zhong K. Alzheimer's disease drug-development pipeline:few candidates, frequent failures. Alzheimers Res. Ther. 2014; 6: 37.

[88] Takeda A, Loveman E, Clegg A, Kirby J, Picot J, Payne E, Green CA. A systematic review of the clinical effectiveness of donepezil, rivastigmine and galantamine on cognition, quality of life and adverse events in Alzheimer's disease. Int. J. Geriatr. Psychiatry. 2006; 21: 17-28.

[89] Honig LS, Vellas B, Woodward M, Boada M, Bullock R, Borrie M, Hager K, Andreasen N, Scarpini E, Liu-Seifert H, Case M, Dean RA, Hake A, Sundell K, Poole Hoffmann V, Carlson C, Khanna R, Mintun M, DeMattos R, Selzler KJ, Siemers E. Trial of Solanezumab for Mild Dementia Due to Alzheimer's Disease. N. Engl. J. Med. 2018; 378: 321-330.

[90] Wojsiat J, Zoltowska KM, Laskowska- Kaszub K, Wojda U. Oxidant/ antioxidant imbalance in Alzheimer's disease: Therapeutic and diagnostic prospects. Oxid. Med. Cell Longev. 2018; 2018: 6435861.

[91] Kim HS, Suh YH. Minocycline and neurodegenerative diseases. Behav. Brain Res. 2009; 196: 168-179.

[92] Ono K, Naiki H, Yamada M. The development of preventives and therapeutics for Alzheimer's disease thatinhibit the formation of beta-amyloid fibrils (fAbeta), as well as destabilize preformed fAbeta. Curr. Pharm. Des. 2006;12: 4357– 4375.

[93] Isopi E, Granzotto A, Corona C, Bomba M, Ciavardelli D, Curcio M, Canzoniero LM, Navarra R, Lattanzio R, Piantelli M, Sensi SL. Pyruvate prevents the development of age-dependent

cognitive deficits in a mouse model of Alzheimer's disease without reducing amyloid and tau pathology. Neurobiol. Dis. 2015; 81: 214-224.

[94] Massey V. The chemical and biological versatility of Riboflavin. Biochem. Soc. Trans. 2000; 28: 283-296.

[95] Bafunno V, Giancaspero TA, Brizio C, Bufano D, Passarella S, Boles E, Barile, M. Riboflavin uptake and FAD synthesis in *Saccharomyces cerevisiae* mitochondria. Involvement of the FLX1p carrier in FAD export. J. Biol. Chem. 2004; 279: 95-102.

[96] Johnson JO, Gibbs JR, Megarbane A, Urtizberea JA, Hernandez DG, Foley AR, Arepalli S, Pandraud A, Sanchez JS, Clayton P, Reilly MM, Muntoni F, Abramzon Y, Houlden H, Singleton AB. Exome sequencing reveals riboflavin transporter mutations as a cause of motor neuron disease. Brain, 2012; 135: 1-8.

[97] Udhayabanu T, Manole A, Rajeshwari M, Varalakshmi P, Houlden H, Ashokkumar B. Riboflavin responsive mitochondrial dysfunction in neurodegenerative diseases. J. Clinic. Medic. 2017; 6: 52.

[98] Coimbra CG, Junqueira VBC. High doses of riboflavin and the elimination of dietary red meat promote the recovery of some motor functions in Parkinson's disease patients. Braz. J. Med. Biol. Res. 2003; 36: 1409-1417.

[99] Naghashpour M, Amani R, Sarkaki A, Ghadiri A, Samarbafzadeh A, Jafarirad S, Saki Malehi A. Brain-derived neurotrophic and immunologic factors: Beneficial effects of riboflavin on motor disability in murine model of multiple sclerosis. Iran. J. Basic Med. Sci. 2016; 19: 439-448.

[100] Bosetti F, Brizzi F, Barogi S, Mancuso M, Siciliano G, Tendi EA, Murri L, Rapoport SI, Solaini G. Cytochrome c oxidase and mitochondrial F1F0-ATPase (ATP synthase) activities in platelets and brain from patients with Alzheimer's disease. Neurobiol. Aging. 2002; 23: 371-376.

[101] Reddy PH, Manczak M, Yin X, Reddy AP. Synergistic Protective Effects of Mitochondrial Division Inhibitor 1 and Mitochondria- Targeted Small Peptide SS31 in Alzheimer's Disease. J. Alzheim. Dis. 2018; 62: 1549-1565.

[102] Zhang W, Gu GJ, Shen X, Zhang Q, Wang GM, Wang PJ. Neural stem cell transplantation enhances mitochondrial biogenesis in a transgenic mouse model of Alzheimer's disease- like pathology. Neurobiol. Aging, 2015; 36: 1282– 1292.

[103] McManus MJ, Murphy MP, Franklin JL. The mitochondria-targeted antioxidant MitoQ prevents loss of spatial memory retention and early neuropathology in a transgenic mouse model of Alzheimer's disease. J. Neurosci. 2011; 31: 15703-15715.

[104] Lim YA, Grimm A, Giese M, Mensah-Nyagan AG, Villafranca JE, Ittner LM, Eckert A, Götz J. Inhibition of the Mitochondrial Enzyme ABAD Restores the Amyloid-B-Mediated Deregulation of Estradiol. PloS One, 2011; 6: e28887.

[105] Du H, Guo L, Fang F, Chen D, Sosunov AA, McKhann GM, Yan Y, Wang C, Zhang H, Molkentin JD, Gunn-Moore FJ, Vonsattel JP, Arancio O, Chen JX, Yan SD. Cyclophilin D deficiency attenuates mitochondrial and neuronal perturbation and ameliorates learning and memory in Alzheimer's disease. Nat. Med. 2008; 14:1097-1105.

[106] Rubinstein JL, Walker JE, Henderson R. Structure of the mitochondrial ATP synthase by electron cryomicroscopy. EMBO J. 2003; 22: 6182-6192.

[107] Beck SJ, Guo L, Phensy A, Tian J, Wang L, Tandon N, Gauba E, Lu L, Pascual JM, Kroener S, Du H. Deregulation of mitochondrial F1FO-ATP synthase via OSCP in Alzheimer's disease. Nat. Commun. 2016; 7: 11483.

[108] Monteiro-Cardoso VF, Oliveira MM, Melo T, Domingues MR, Moreira PI, Ferreiro E, Peixoto F, Videira RA. Cardiolipin profile changes are associated to the early synaptic mitochondrial dysfunction in Alzheimer's disease. J. Alzheimer's Dis. 2015; 43: 1375-1392.

[109] Romani M. Enhancing mitochondrial quality control to fight neuromuscular degeneration in aging and disease. Thesis, Lausanne EPFL, 2021.

[110] Gupta VK, Scheunemann L, Eisenberg T, Mertel S, Bhukel A, Koemans TS, Kramer JM, Liu KS, Schroeder S, Stunnenberg HG, Sinner F, Magnes C, Pieber TR, Dipt S, Fiala A, Schenck A, Schwaerzel M, Madeo F, Sigrist SJ. Restoring polyamines protects from age-induced memory impairment in an autophagy-dependent manner. Nat. Neurosci. 2013; 16:1453-1460.

[111] Turunc Bayrakdar E, Uyanikgil Y, Kanit L, Koylu E, Yalcin A. Nicotinamide treatment reduces the levels of oxidative stress, apoptosis, and PARP-1 activity in Abeta(1-42)-induced rat model of Alzheimer's disease. Free Radic. Res. 2014; 48:146-158.

[112] Geisler JG, Marosi K, Halpern J, Mattson MP. DNP, mitochondrial uncoupling, and neuroprotection: A little dab'll do ya. Alzheimers Dement. 2017; 13: 582-591.

[113] Qian C, Yu YJ. Mitophagy in Alzheimer's Disease and Other Age-Related Neurodegenerative Diseases Cells. 2020; 9: 150; doi: 10.3390 /cells 9010150.

[114] Lipinski MM, Zheng B, Lu T, Yan ZY, Py BF, Ng A, Xavier RJ, Li C, Yankner BA, Scherzer CR, Yuan JY. Genome-wide analysis reveals mechanisms modulating autophagy in normal brain aging and in Alzheimer's disease. Proc. Natl. Acad. Sci. U.S.A. 2010; 107: 14164-14169.

[115] Dusetti NJ, Jiang Y, Vaccaro MI, Tomasini R, Azizi Samir A, Calvo EL, Ropolo A, Fiedler F, Mallo GV, Dagorn JC, Iovanna JL. Cloning and expression of the rat vacuole membrane protein 1 (VMP1), a new gene activated in pancreas with acute pancreatitis, which promotes vacuole formation. Biochem. Biophys. Res. Commun. 2002; 290;641-649.

[116] Wang P, Xi Chen Y, Wang CJ, Liu X, Wang Y, Wu H, Cai H, Han-Ming Shen Le W. Essential role for autophagy protein VMP1 in maintaining neuronal homeostasis and preventing axonal degeneration. Cell Death and Disease. 2021; 12:116-130.

[117] Komatsu M, Wang QJ, Holstein GR, Friedrich VL Jr, Iwata J, Kominami E, Chait BT, Tanaka K, Yue Z. Essential role for autophagy protein Atg7 in the maintenance of axonal homeostasis and the prevention of axonal degeneration. Proc. Natl. Acad. Sci. USA. 2007; 104: 14489-14494.

[118] Brines M, Cerami A. Emerging biological roles for erythropoietin in the nervous system. Nat. Rev. Neurosci. 2005; 6: 484-494.

[119] Jang W, Park J, Shin KJ, Kim JS, Youn J, Cho JW, Oh E, Ahn JY, Oh KW, Kim HT. Safety and efficacy of recombinant human erythropoietin treatment of non-motor symptoms in Parkinson's disease. J. Neurol. Sci. 2014; 337: 47-54.

[120] Rey F, Ottolenghi S, Giallongo T, Balsari A, Martinelli C, Rey R, Allevi R,

Giulio AMD, Zuccotti GV, Mazzucchelli S, Foresti R, Samaja M, Carelli S. Mitochondrial Metabolism as Target of the Neuroprotective Role of Erythropoietin in Parkinson's Disease. Antioxidants (Basel). 2021;10:121.

[121] Copeland DE, Dalton AJ. An association between mitochondria and the endoplasmic reticulum in cells of the pseudobranch gland of a teleost. J. Biophys. Biochem. Cytol. 1959; 5: 393-396.

[122] Bereiter-Hahn J. Behavior of mitochondria in the living cell. Int. Rev. Cytol. 1990; 122: 1-63.

[123] Area-Gomez E, de Groof AJ, Boldogh I, Bird TD, Gibson GE, Koehler CM, Yu WH, Duff KE, Yaffe MP, Pon LA, Schon EA. Presenilins are enriched in endoplasmic reticulum membranes associated with mitochondria. Am. J. Pathol. 2009; 175: 1810-1816.

[124] Weiwei Y, Haiqiang J, Yining H. Mitochondria-associated membranes (MAMs): a potential therapeutic target for treating Alzheimer's disease. Clinical Science. 2021; 135: 109-126.

[125] García-Pérez C, Hajnóczky G, Csordás G. Physical coupling supports the local Ca2+ transfer between sarcoplasmic reticulum subdomains and the mitochondria in heart muscle. J. Biol. Chem. 2008; 283: 32771-32780.

[126] Wang X, Zheng W. Ca2+ homeostasis dysregulation in Alzheimer's disease: a focus on plasma membrane and cell organelles. FASEB J. 2019; 33: 6697-6712.

[127] Flis VV, Daum G. Lipid transport between the endoplasmic reticulum and mitochondria. Cold Spring Harb, Perspect, Biol. 2013; 5: a013235 doi: 10. 1101/ cshperspect. a013235.

[128] Sheu SS, Nauduri D, Anders MW. Targeting antioxidants to mitochondria: a new therapeutic direction. Biochim. Biophys. Acta, 2006; 1762: 256-265.

[129] Zhao K, Zhao GM, Wu D, Soong Y, Birk AV, Schiller PW, Szeto HH. Cell-permeable peptide antioxidants targeted to inner mitochondrial membrane inhibit mitochondrial swelling, oxidative cell death, and reperfusion injury. J. Biol. Chem. 2004; 279: 34682-34690.

[130] Hemachandra RP, Manczak M, Kandimalla R. Mitochondria-targeted small molecule SS31: a potential candidate for the treatment of Alzheimer's disease. Hum. Mo. Gen. 2017; 26: 1483-1496.

[131] Calkins MJ, Manczak M, Reddy PH. Mitochondria-targeted antioxidant SS31 prevents amyloid beta-induced mitochondrial abnormalities and synaptic degeneration in Alzheimer's disease. Pharmaceuticals (Basel), 2012; 5: 1103-1119.

[132] Li PA, Hou X, Hao S. Mitochondrial biogenesis in neurodegeneration. J. Neurosci. Res. 2017; 95: 2025-2029.

[133] Sheng B, Wang X, Su B, Lee HG, Casadesus G, Perry G, Zhu X. Impaired mitochondrial biogenesis contributes to mitochondrial dysfunction in Alzheimer's disease. J. Neurochem. 2012; 120: 419-429.

[134] Uittenbogaard M, Chiaramello A. Mitochondrial biogenesis: A therapeutic target for neurodevelopmental disorders and neurodegenerative diseases. Curr. Pharm. Des. 2014; 20: 5574-5593.

[135] Gong B, Pan Y, Vempati P, Zhao W, Knable L, Ho L, Wang J, Sastre M, Ono K, Sauve AA, Pasinetti GM. Nicotinamide riboside restores cognition through an upregulation of proliferator-activated receptor-γ

coactivator 1α regulated β-secretase 1 degradation and mitochondrial gene expression in Alzheimer's mouse models. Neurobiol. Aging. 2013; 34: 1581-1588.

[136] Reddy PH, Manczak M, Yin X, Grady MC, Mitchell A, Kandimalla R, Kuruva CS. Protective effects of a natural product, curcumin, against amyloid β induced mitochondrial and synaptic toxicities in Alzheimer's disease. J. Investig. Med. 2016; 64: 1220-1234.

[137] Heneka MT, Fink A, Doblhammer G. Effect of pioglitazone medication on the incidence of dementia. Ann. Neurol. 2015; 78: 284-294.

[138] Watson GS, Cholerton BA, Reger MA, Baker LD, Plymate SR, Asthana S, Fishel MA, Kulstad JJ, Green PS, Cook DG, Kahn SE, Keeling ML, Craft S. Preserved cognition in patients with early Alzheimer disease and amnestic mild cognitive impairment during treatment with rosiglitazone: a preliminary study. Am. J. Geriatr. Psychiatry. 2005; 13: 950-958.

[139] Wang CF, Song CY, Wang X, Huang LY, Ding M, Yang H, Wang P, Xu LL, Xie ZH, Bi JZ. Protective effects of melatonin on mitochondrial biogenesis and mitochondrial structure and function in the HEK293-APPswe cell model of Alzheimer's disease. Eur. Rev. Med. Pharmacol. Sci. 2019; 23: 3542-3550.

[140] Shukla M, Govitrapong P, Boontem P, Reiter RJ, Satayavivad J. Mechanisms of Melatonin in Alleviating Alzheimer ' s Disease. Curr. Neuropharmacol. 2017;15:1010-1031.

[141] Espino J, Bejarano I, Redondo PC, Rosado JA, Barriga C, Reiter RJ, Pariente JA, Rodríguez AB. Melatonin reduces apoptosis induced by calcium signaling in human leukocytes: Evidence for the involvement of mitochondria and Bax activation. J. Membr. Biol. 2010; 233:105-118.

[142] Mladinov M, Yeop Oh J, Petersen C, Eser R, Hua Li S, Theofilas P, Spina S, Seeley WW, Bittencourt JC, Neylan TC, Grinberg LT. Specific pattern of melanin-concentrating hormone (MCH) neuron degeneration in Alzheimer's disease and possible clinical implications medRxiv 2021;01.27.21250608; doi: https: //doi. org/ 10.1101 /2021. 01. 27.21250608

[143] Hashemi M, Ghavami S, Eshraghi M, Booy EP, Los M. Cytotoxic effects of intra and extracellular zinc chelation on human breast cancer cells. Eur. J. Pharmacol. 2007; 557 : 9-19.

[144] de la Torre JC, Mussivand T. Can disturbed brain microcirculation cause Alzheimer's disease? Neurol. Res. 1993; 15: 146-153.

[145] Zhao Y, Gong CX. From chronic cerebral hypoperfusion to Alzheimer-like brain pathology and neurodegeneration. Cell Mol. Neurobiol. 2015; 35: 101-110.

[146] Govindpani K, McNamara LG, Smith NR, Vinnakota C, Waldvogel HJ, Faull RL, Kwakowsky A. Vascular dysfunction in Alzheimer's disease: A prelude to the pathological process or a consequence of it? J. Clin. Med. 2019; 8: 651.

[147] de la Torre JC. Deciphering Alzheimer's Disease Pathogenic Pathway: Role of Chronic Brain Hypoperfusion on p-Tau and mTOR. J. Alzheimers Dis. 2021; 79: 1381-1396.

[148] MacAskill AF, Kittler JT. Control of mitochondrial transport and localization in neurons. Trends Cell Biol. 2010; 20: 102-112.

[149] Kim HA, Miller AA, Drummond GR, Thrift AG, Arumugam TV, Phan TG, Srikanth VK, Sobey CG. Vascular cognitive impairment and Alzheimer's disease: Role of cerebral hypoperfusion and oxidative stress. Naunyn Schmiedebergs Arch. Pharmacol. 2012; 385: 953-959.

[150] Marco M, Shaw PJ, Ferraiuolo L, Blackburn DJ, Venneri A, Mortiboys H. Mitochondrial Dysfunction in Alzheimer's Disease: A Biomarker of the Future? Biomedicines 2021; 9: 63. https://doi.org/10.3390/biomedicines9010063

Chapter 6

Reversal of Cognitive Aging through Enhancement of Cardiac Output

Kenneth J. McLeod

Abstract

Cognitive aging is a progressive condition leading to dementia, a condition which is now the sixth leading cause of death in the U.S., as well as being among the most expensive healthcare conditions to manage. With over 5 million affected in the U.S. alone, the annual costs to the Medicare/Medicaid system exceeds $200 billion, and with the rising age of the population, annual costs of dementia care are expected to exceed $500 billion by 2040. As there is no cure for dementia, a consensus has formed that a more pragmatic goal of research should be developing interventions capable of slowing or preventing cognitive aging. We propose that this is a readily achievable goal. Cognitive impairment is closely linked to cerebral perfusion, and cerebral perfusion is a function of cardiac output. In turn, cardiac output is completely dependent on venous return, which in the upright human, relies on adequate soleus muscle activity. As modern adults rarely squat, which is necessary for maintaining the soleus muscle, soleus insufficiency develops early in adulthood in most people. However, soleus muscle insufficiency can be reversed, resulting in improved cardiac output, cerebral perfusion, and the prevention of cognitive aging.

Keywords: dementia prevention, cerebral perfusion, cardiac output, venous return, second heart, soleus muscle stimulation

1. Introduction

Dementia is a major cause of morbidity and mortality in the developed world. Dementia, in all of its forms, is a progressive condition, with an incidence of less than 5% through age 79, but reaching 40% for those over age 90 [1]. Given the aging demographics of the developed world, the economic impact of this condition could soon dominant healthcare costs in many countries.

There is currently no cure for dementia, and numerous pharmaceutical firms have abandoned the search for a cure. In particular, interventions based on the beta-amyloid hypothesis which has guided dementia drug therapy development for the last three decades has come under increasing scrutiny as drugs which effectively reduce beta-amyloid accumulation appear to exacerbate, rather than ameliorate, the symptoms associated with dementia [2]. It is therefore incumbent that we take a fresh approach to understanding dementia, in particular, we suggest it is important to develop a more thorough understanding of the numerous physiologic interactions associated with progression of cognitive impairment with age. Such understanding will set the stage for innovative interventions, specifically, interventions focused on prevention, rather than treatment. This coupled systems, or complex systems, approach, is less intuitive than the more traditional scientific approach of establishing proximate cause. Indeed, in complex systems, cause may not be identifiable, rather, outcomes arise as emergent behaviors of interdependent coupled components of the system. Despite these challenges, it is becoming widely recognized that a complex systems mindset will be necessary for effectively addressing not only dementia, but also the wide range of functional disorders which modern medicine currently faces [3].

Perhaps the physiologic interactions of greatest current interest, with respect to dementia, are those between the cardiovascular and cerebral systems. Over the past three decades, numerous prospective and retrospective studies have identified strong associations between low cardiac output, low blood pressure, low cerebral perfusion, and the development of dementia. The majority of these studies have focused on older adults, but a review of cognitive and cardiovascular changes taking place from early adulthood provides important insights into why dementia may not have to be the scourge of old age which many people fear.

Here, we describe the development of cognitive decline starting in early adulthood and relate this decline to parallel changes in the cardiovascular and the musculo-skeletal systems, specifically, second heart function. The parallels in secular decline in these systems lead us to propose that inactivity based changes in skeletal muscle fiber structure plays a critical role in the age related decline in cardiac output, and correspondingly decreased cerebral blood flow. We propose that this decreased cerebral blood flow, beginning in middle age, is a dominant factor in cognitive decline, cognitive impairment, and eventually dementia, in those where cerebral perfusion is not corrected. We introduce preliminary evidence showing that enhancement of cardiac output through second heart (soleus muscle) stimulation is able to improve cognitive performance in those with both mild and advanced cognitive impairment.

2. Age related cognitive decline

Dementia is a syndrome characterized by memory loss, decline in executive function, behavioral changes, and ability to perform activities of daily living. The impact of dementia on the healthcare system is by far the highest of any health condition [4]. In the U.S., for example, cumulative five year care costs exceed $300,000 or roughly twice the cost of care for heart disease or cancer. With almost 6 million Americans currently affected, annual costs to Medicare/Medicaid exceed $300 billion, and with the aging of the population, these costs are expected to exceed $500 billion by 2040, unless an effective intervention is developed.

While prevalence within the elderly population is based on diagnosed cases, it has become clear that dementia is a slowly progressive condition initiated at a far earlier stage of life. For example, in our laboratory, we have utilized computer aided assessments (Cognivue, Inc., Victor, NY) to quantify cognitive function, including memory, motor skills, and executive function. We have observed (**Figure 1**) that by age 65, cognitive performance for more than 50% of individuals falls below established threshold for mild cognitive impairment. Moreover, by age 80, roughly one-half of the individuals we have screened in our laboratory score below the threshold for moderate to severe cognitive impairment.

Remarkably, we observe few individuals over the age of 55 who are able to score above 90 on the Cognivue scale (scores above 95 are readily attained by young adults). Linear regression leads to the suggestion that cognitive decline is initiated while individuals are still in their 30s. This perspective is confirmed by the work of Hughes et al. who have investigated cognitive performance among middle-aged and older individuals through the use of telephone-based assessments [5]. In assessing over 2500 individuals using a range of validated assessments, small declines in cognitive performance were observed as people progressed from their 30s into the 40s, however, only one assessment (backwards counting) showed a significant decline in performance over this decade. Starting in the 40s, dramatic declines became evident relative to that of individuals in their 30s (**Figure 2**).

The characteristics of cognitive performance decline appear to be dependent on cognitive task. Short-term memory skills, such as repeating a digit sequence

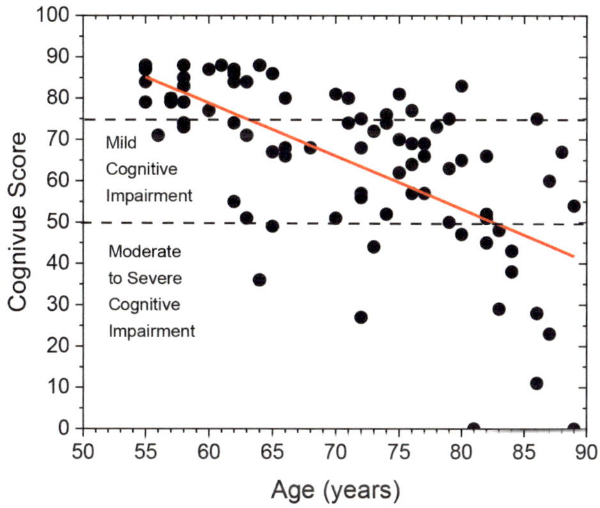

Figure 1.
Age related decline in cognitive performance (memory, motor, and executive function) in a convenience sample of middle aged and older subjects. By age 65, more than half of tested subject perform at a level characterized as mild cognitive impairment or worse. For those in their mid 80s or older, more than half perform at a moderate to severe cognitive impairment level.

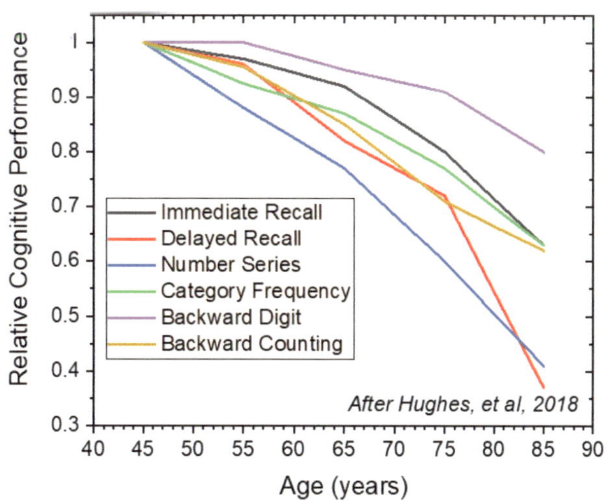

Figure 2.
Decline in cognitive function in middle aged adults. In phone evaluations of over 2500 middle aged and older adults, significant decline in backward counting capability becomes evident between 30 and 40 years of age. Beyond age 40, the majority of cognitive skills are found to decline, and beyond age 50, all cognitive skills evaluated decline with increasing age. After Hughes, et al., [5].

backward, declined slowly with age, along with immediate recall tasks. However, executive function tasks were found to already show substantial decline in early middle age.

3. Age related decline in cardiac output and blood pressure

The question naturally arises as to whether these observed "declines" in cognitive performance in middle-age individuals are, in fact, detrimental, or instead simply reflect more effective use of "brain power" which comes with experience. That is, individuals consistently show improved performance in day to day functions in this age range despite the decline observed in cognitive testing [6]. A generally accepted explanation for this apparent paradox is that, for young people, most daily experiences are novel, and so they retain a high ability to deal with novel exposures. Alternatively, by middle age, most people have obtained a knowledge base of serviceable answers to the most commonly encountered mental challenges, which they can recover with minimal cognitive effort. Because cognitive testing, by design, relies on the presentation of novel challenges, this gives young individuals a natural advantage independent of actual levels of cognitive capability. Certainly, many individuals retain "normal" levels of cognitive performance as measured by cognitive testing well into their 80s, as seen in **Figure 1**, perhaps indicating that these individuals have retained the ability to deal with fresh challenges through regular exposure to novel experiences.

Nonetheless, the consensus in the healthcare community is that while dementia is not a normal outcome of aging, some cognitive decline is to be expected with aging. Age related changes in cardiovascular system performance provides a physiologic basis for this consensus. Specifically, cardiac output has long been observed to decline with age. However, early demonstrations of this declining pattern have relied on invasive measurement techniques which were capable of creating a stress response which may have influenced these older measurements. To address this issue, Middlemiss, et al. [7] have recently utilized non-invasive cardiac output assessment techniques to evaluate changes in cardiac output across the adult age span (**Figure 3**).

These non-invasive measures confirm that cardiac output in the supine position declines substantially with age, specifically by almost 50% over the adult life span in both men and women. Moreover, these investigation shows that cardiac output declines by a further 25% during transition from the supine to the seated position, and falls by 50% when transitioning from supine to a standing position. The implication is that cardiac output in older adults who are standing quietly can be expected to be reduced, on average, by 75% in comparison to that of an average 20 year old.

Cardiac output is a key determinant of arterial blood pressure. In combination with peripheral vascular resistance, cardiac output establishes mean blood pressure. As sufficient blood pressure must be sustained throughout the cardiac cycle in order to ensure adequate blood flow to the brain, which is located at the top of the body when in upright posture, blood pressure becomes a critical factor in regulating cognitive performance. In principle, the declining cardiac output associated with aging should not necessarily lead to declining blood pressure, as vasoconstriction can raise peripheral vascular resistance in order to maintain blood pressure levels. In fact, given the dramatic decline in cardiac output when upright, in the majority of older individuals the ability to vaso-constrict is insufficient to maintain normal blood pressure.

In our lab, we focus on assessing resting diastolic blood pressure (DBP), as the lowest pressure during the diastolic phase of the heart contraction cycle represents the point at which cerebral blood flow is at a minimum. We obtain resting DBP with the subjects in a quiet, seated position for at least 10 minutes, and record the third of three brachial pressure measurements. We observe (**Figure 4**) that by age 55, average resting DBP is below 80 mmHg. By the 9th decade of life, average diastolic pressures are below 70 mmHg. Overall, we observe that among this convenience sample that approximately 20% are unable to maintain a resting diastolic pressure above 65 mmHg, a level at which symptoms of orthostatic hypotension (OH) become evident. For subjects over the age of 75, 30% are unable to maintain this threshold DBP level.

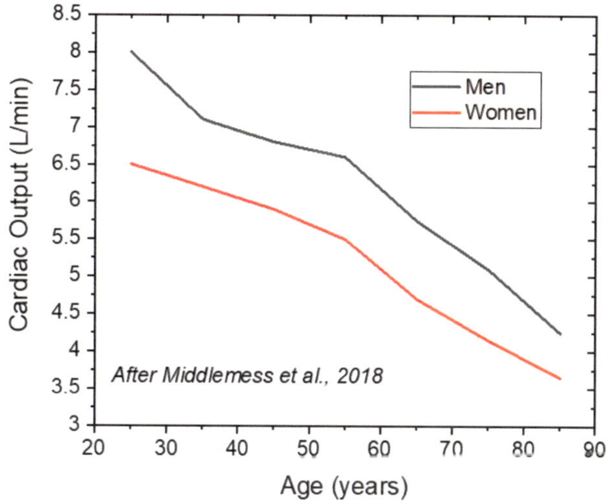

Figure 3.
Cardiac output (CO) as a function of age as measured utilizing non-invasive assessments. Cardiac output declines by approximately 50% in both men and women from the 3rd decade to 9th decade of life. CO decline occur at a relatively constant rate over this age range.

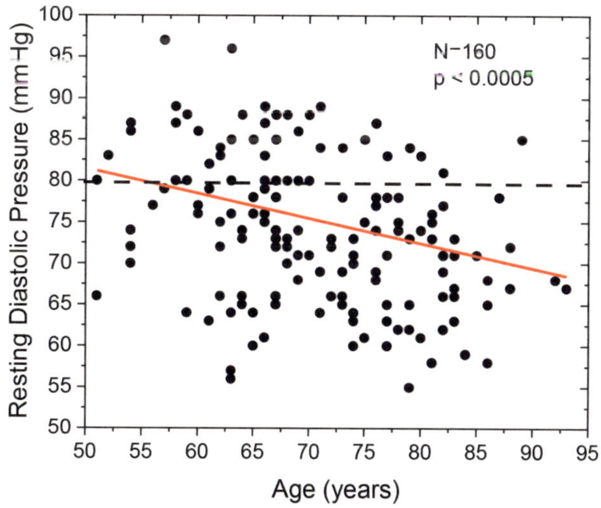

Figure 4.
Resting diastolic blood pressure (DBP) vs. age. A robust negative correlation ($p < 0.0005$) is observed between DBP and age with average DBP falling below 80 mmHg by age 55 in a convenience sample of men and women. By age 85, average DBP falls below 70 mmHg, and within the subject population, approximately 20% were unable to sustain a DBP above 65 mmHg after sitting for 10 minutes.

OH has been shown to occur in less than 3% of young adults, but up to 35% in individuals over the age of 75 [8]. Torabi, et al. [9] investigated the cardiovascular characteristics of individuals with both classical and delayed OH. In a study of over 2000 patients, over the age of 15, with unexplained syncope, 27% were found to be unable to maintain normal blood pressure levels during upright tilt testing. In this population, systolic blood pressure fell, on average to 95 mmHg, while diastolic blood pressure fell, on average, to 60 mmHg.

These observations indicate that asymptomatic postural hypotension is remarkably common in the adult population. That is, vaso-constrictive ability is insufficient in at least 20% of the adult population to maintain normal blood pressure during quiet sitting. Moreover, among the older population, symptomatic postural hypotension is evident in over one-third of individuals, who are unable to maintain blood pressure levels in the presence of declining cardiac output. The critical question is whether the health implications associated with chronic hypotension extend beyond the inconveniences of dizziness and occasional syncope. Extensive work on the association of hypotension with cognitive impairment suggests that hypotension, and correspondingly, cerebral hypo-perfusion, may be one of the most consistent risk factors associated with dementia.

4. Role of cerebral perfusion in cognitive function

Numerous lines of evidence lend strong support for the hypothesis that sustained cerebral hypo-perfusion as a result of chronically low blood pressure has significant negative effects on cognitive performance, and as well, leads to the development of dementia.

Recent computer aided cognitive assessments of men and women over the age of 50, for example, demonstrate a strong correlation between resting diastolic pressure and cognitive performance (**Figure 5**). Multivariate regression analysis on these data show that, after adjusting for subject age, resting diastolic pressure is a significant ($p < 0.02$) predictor of cognitive performance with close to a 1% decline in performance for each 1 mmHg drop in DBP. Notably, only for average diastolic blood pressures above 80, is normal cognitive performance (assessment score > 75) observed. Similarly, the regression analysis indicates that for diastolic pressures below 50 mmHg, average cognitive performance falls into the moderate cognitive impairment range.

These results are consistent with those first reported in the Baltimore Longitudinal Aging Study [10] where it was found that cognitive performance in an older (70 ± 8 years) population was significantly degraded at diastolic blood pressures below 80 mmHg. Confirmation is also obtained by comparison of age dependent cardiovascular and cognitive performance measures (**Figure 6**). Combining the results of Middlemiss et al. [7] with the results of Hughes et al. [5] demonstrates a robust ($p = 0.002$) association between cognitive performance and cardiac output. This analysis demonstrates that for a 30% decline in cardiac output, a 40% decline in cardiac performance can be expected in the 40–90 year old population.

Over the past two decades, numerous studies have provided substantial evidence that decreased cardiac output and chronically low blood pressure are associated with declines in cognitive performance, and also significantly increases the risk of developing dementia. Among the earliest of these studies was the Kugsholmen project undertaken in Sweden [11]. This study showed that, in an elderly population, those with a systolic blood pressure below 140 mmHg, or a diastolic blood pressure below 75 mmHg, had a 3x greater likelihood of being diagnosed with dementia. At that point in time it was unclear whether the lower blood pressures were a consequence of dementia, or played a causal role.

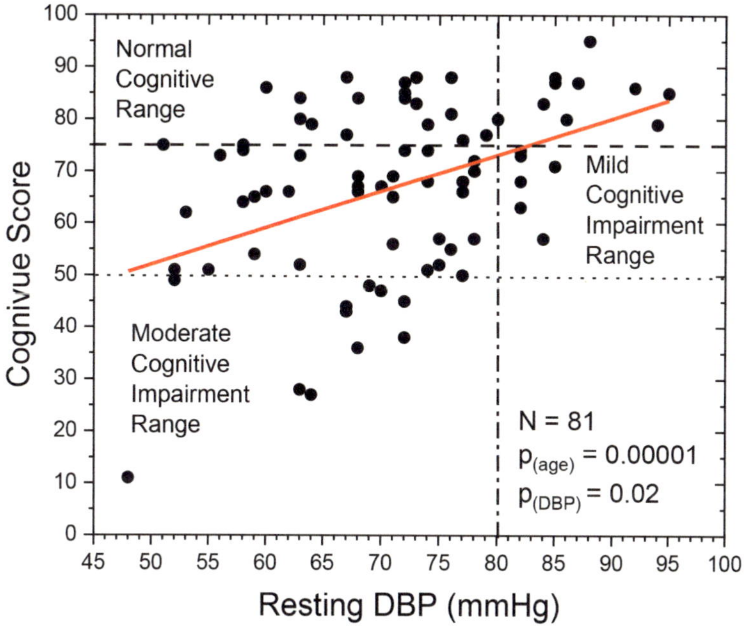

Figure 5.
Cognitive performance vs. resting diastolic blood pressure (DBP). After adjusting for age effects on cognition, declining DBP is strongly associated with declining cognition levels in a convenience sample of men and women. The average individual with a resting DBP below 80 mmHg falls into the category of mild cognitive impairment as assessed using the computer aided Cognivue assessment. DPB below approximately 50 mmHg is associated with transition into the range of moderate to severe cognitive impairment.

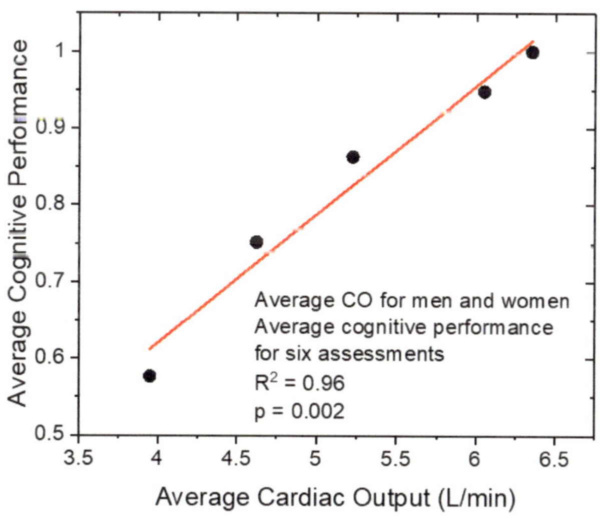

Figure 6.
Integrated analysis of age related cognitive performance data per Hughes et al. [5] with age related cardiac output data per Middlemiss et al. [7]. Cardiac output robustly predicts (p = 0.002) cognitive performance, consistent with a causal role for decreased cerebral blood flow mediating cognitive decline.

The East Boston study [12] addressed, in part, this question, by showing that there was an inverse correlation between risk of Alzheimer's diagnosis and blood pressures taken four years before diagnosis. Verghese, et al. [13] subsequently directly addressed this question in the Bronx Aging Study. They observed in this community based, longitudinal study that sustained low diastolic blood pressure (<70 mmHg) was associated with a 2x increased risk of developing Alzheimer's disease over a 20 year period. More recently, in a cross-sectional study of more than 24,000 adults who did not have a dementia diagnosis, subjects were followed for up to 27 years [14]. Applying multiple linear regression to adjust for age, gender, education, and body mass index, significant negative correlations were observed between risk of developing Alzheimers, as well as all-cause dementia, across the full range of systolic and diastolic blood pressures.

Complementing these investigation, the established link between diabetes and risk of dementia [15], combined with the well-known influences of diabetes on vascular dysfunction, is currently leading to a broader acceptance that hypo-metabolism, and correspondingly, hypo-perfusion, plays a more significant role in the development of dementia than previously considered [16].

5. Mechanism of age related decline in cardiac output

The numerous demonstrated associations between declines in cardiac output, blood pressure, cognitive performance, and risk of developing dementia, provides a physiologic explanation for the age related cognitive decline, but provides limited insight into how this decline could be prevented or reversed. Our research has led us to propose that the critical factor linking these related outcomes is the inability to maintain adequate venous return during orthostasis.

Venous return refers to the flow of blood from the periphery of the body back to the right atrium. While venous return and cardiac output levels can transiently deviate, under normal physiologic conditions cardiac output is strictly a function of venous return. In the supine position, venous resistance contributes only about 15% to total vascular resistance, however, in upright posture, the venous system plays a much larger role in influencing venous return.

The largest influence of the venous system is through its role as a capacitance vessel. Veins are highly distensible, having thinner walls, with larger diameters, and a compliance of about 30 times that of arteries. They can, therefore, expand rapidly to accommodate large volumes of blood. Correspondingly, a transition from supine to upright posture typically leads to a rapid 500 ml redistribution of blood to the peripheral venous system, a fluid shift which continues to increase over time. The ratio of venous to arterial capacitance under orthostasis has been estimated to grow to as large as 18:1 [17].

In addition, the influence of gravity on the hydrostatic column of blood in the venous system is such that venous blood pressure in the feet can exceed 90 mmHg. As a result of these high lower limb pressures, fluid extravasation from the vascular system increases. Increased extravasation can lead to an additional loss of up to 750 ml over 30–40 minutes following the transition to upright posture. Not only does this cause a further decrease in circulatory system blood volume, but also increased interstitial fluid pressure which results in compression of the peripheral vasculature, and increased in vascular resistance.

The net effect of reduced circulatory volume and increased vascular resistance during upright posture is significantly decreased cardiac output. While vaso-constriction serves to partially support blood pressure during orthostasis, this additional increase in vascular resistance also serves to further reduce blood flow. In our lab, we have observed average sustained decreases in cardiac index, resulting from a transition to quiet sitting, of over 35% relative to that supported when individuals were supine (**Figure** 7).

Return of pooled blood and interstitial fluid which occurs during orthostasis is critically dependent on skeletal muscle pumping. While locomotion can play a role in this process, most adults are sedentary for 9–10 hours per day [18]. Under

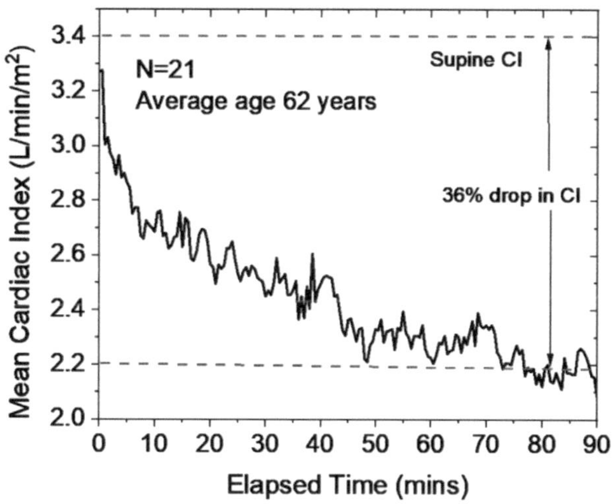

Figure 7.
Cardiac index as a function of time following transition from supine to upright sitting. A decline in CI of 36% from a supine CI of 3.4 L/min/m² is observed among healthy adult women with an average age of 62. This occurs despite an increase in metabolic rate associated with an upright posture, and arises due to gravity induced blood pooling in the lower exptremities.

sedentary conditions, skeletal muscle pumping activity is dominated by soleus muscle action. The essential role played by the soleus in ensuring venous and lymphatic return during orthostasis has led to these muscles commonly being referred to as the calf muscle pumps, or the "second hearts."

The soleus muscles are highly specialized muscles which contain up to 18 thin walled sinuses, each of which are able to hold large volumes of blood. Further, as deep postural muscles composed primarily of slow-twitch fibers, the soleus muscles can sustain contractions over extended time periods. A typical soleus contraction cycle lasts up to one minute, followed a relaxation phase of 60–90 seconds during which the sinuses are able to refill. In addition, because the soleus muscles originate on the posterior tibia and fibula, the muscle can pump effectively when a person is seated. These muscles can generate venous driving forces exceeding 200 mmHg, more than sufficient to drive blood and interstitial fluid back to the heart during upright posture.

6. Soleus muscle adaptation

Like all muscle tissues, the soleus muscle demonstrate changes in both structure and physiology with increasing age. The most commonly observed change in voluntary muscle with advancing age is reduction in muscle mass, with Type II muscle fibers decreasing in both numbers and in volume with age [19]. However the soleus muscle is a deep postural muscle and principally composed of Type I fibers, and Type I fibers do not change substantially in size or number with advancing age. Rather, Type I fibers are far more affected by usage patterns.

Specifically, lack of use of the soleus muscle results in fibers converting towards Type II behavior. Microvascular supply to the fibers is lost and correspondingly, the innate fatigue resistance expected in deep postural muscle tissue. This transition can occur rapidly, independent of age. NASA studies characterizing muscle fiber type changes in astronauts found more than a 20% loss in force generating capacity in both Type I and Type IIa fibers taken from the soleus after a remarkably short (17 day) space flight [20].

The postural role of the soleus muscle is plantar flexion. In fact, when an individual is in a bent knee position, the soleus is the only active plantar flexion muscle. The postural activities which require the most significant plantar flexion force in the bent knee position are squatting activities. Squatting is the natural human rest position, and our ancestors squatted regularly throughout the day - while cooking, eating, socializing, and of course, when defecating. Children also commonly squat during the day, but in the modern world, sitting has become the dominant resting position. While a small level of soleus activity occurs during sitting, squatting results in 4–5 times as much soleus muscle activity as sitting [21]. Therefore, while our ancestors were typically sedentary for 9 or more hours each day, similar to modern individuals, their natural resting posture required up to 5x more soleus muscle activity, thereby persevering the fatigue resistant qualities of the slow twitch muscle fibers in this muscle.

7. Soleus muscle stimulation

The critical observation is that the commonly observed declines in the venous return of adults is not a function of age, per se, but rather is the result of the transition to sitting as a dominant resting posture, in particular as people get older. The transition to sitting as the dominant upright resting posture for adults has resulted in two significant impacts on venous return. First, the soleus muscles are activated for only a small fraction of the time when people are sedentary, and correspondingly, muscle pump activity is limited. Second, soleus inactivity results in an adaptation of the soleus muscle fibers such that the muscle, even when activated, is unable to develop the sustained forces necessary to ensure adequate venous return.

Because the soleus fiber adaptations which occur in most people arise primarily from disuse and not due to aging, reconversion of the soleus muscle fibers back to Type I fibers should be possible through alteration of muscle activation patterns. The soleus muscles are activated, when in upright posture, when the center of gravity of the body moves too far forward; soleus contraction returns the body to a balanced position. This shift in the center of gravity is sensed by pressure on the frontal plantar surface, specifically by Meissner's Corpuscles, which activate short, and long, loop reflex arcs which trigger soleus contraction.

Retraining of the soleus muscle fibers therefore should simply require a sustained stimulation of the postural reflex arc in a pattern which mimics normal resting posture (i.e. squatting) activation. This can be achieved using micromechanical stimulation of the Meissner's Corpuscles periodically for sustained periods of time (one minute bouts) for extended time periods, over the course of the day (i.e. a significant fraction of sedentary time).

In our lab, we have undertaken such studies utilizing the soleus muscle stimulator (HeartPartner) developed by Sonostics, Inc. (Endicott, NY). We utilize electrical impedance plethysmography (Cheetah Medical; Wilmington, DE) to track cardiac output following a transition from the quiet standing position to quiet sitting. This represents a change in metabolic activity from about 1.74 METS to about 1.46 METS [22] or roughly a 17% decline in metabolic demands and therefore cardiac output (CO). Typically, the decline observed in adults in far greater. **Figure 8** provides an example of the observed cardiac performance in response to this shift in posture in an older adult. From an initial cardiac index (CI=CO/Body Surface Area) of 2.8 L/min/m^2, cardiac output drops by almost 40% during 60 minutes of quiet sitting.

These results demonstrate that, when seated, the soleus muscles are commonly not being stimulated sufficiently to sustain the venous return necessary to maintain

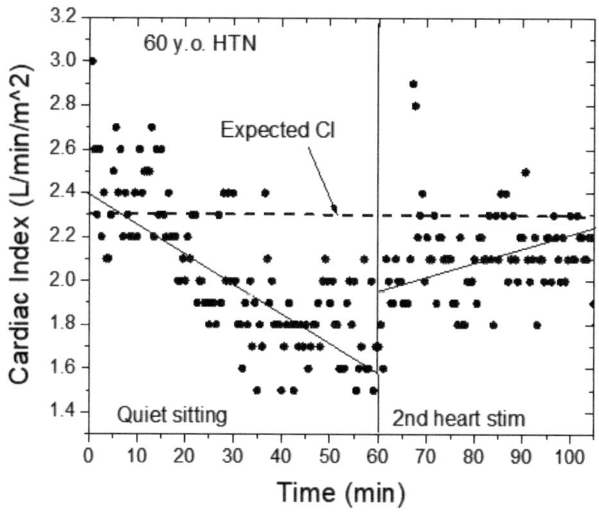

Figure 8.
Cardiovascular system response to soleus muscle stimulation in a 60 year old woman. A change in posture from standing to sitting results in this individual results in a 40% decline in cardiac index (CO/BSA) whereas 20% or less would be expected. While quiet sitting is incapable of stimulate the soleus muscles sufficiently to maintain the venous return necessary to sustain a normal level of cardiac output, external stimulation of the soleus muscles is seen to be capable of returning cardiac output to normal levels within 30 minutes. Initial abrupt rise in CI reflects return of blood pooled into lower limb veins, while the slower rise in CI reflects interstitial fluid return through the lymphatics.

normal cardiac output (CO). However, the soleus muscles still respond, at least over a relatively short duration (30 minutes), to external stimulation. The initial (within minutes) response to soleus stimulation is a rapid rise in cardiac output due to the return of blood pooled into the lower leg veins. Over tens of minutes, interstitial fluid return through the lower limb lymphatics serves to further increase cardiovascular volume resulting in a return to a cardiac output level expected for a sitting adult.

Importantly, just as the soleus muscle rapidly adapts to disuse, these muscles appear to be capable of rapidly "readapting" or more specifically, undergoing muscle fiber reconversion. **Figure 9** (left panel) illustrates the cardiovascular response to the orthostatic stress of quiet sitting in a young (35 year old) woman with severe second heart insufficiency. Upon transitioning from a standing to a sitting position, venous return is inadequate to maintain resting diastolic pressure above a hypotensive level. Specifically, following a transition from standing to quiet sitting, her diastolic pressure is seen to decline from about 80 mmHg, to less than 55 mmHg. Though sitting provides insufficient stimulation to the soleus muscles to maintain venous return, her soleus muscles remain capable of responding to external stimulation. Sustained soleus stimulation over 30 minutes returns her diastolic pressure back close to the normal range (~75 mmHg).

Three months of daily soleus muscle stimulation, for at least one hour per day, resulted in a substantially improved cardiovascular response to the orthostatic stress of quiet sitting in this subject (**Figure 9** right panel). While sitting still resulted in a drop in diastolic blood pressure, the decline it seen to occur at a much slower rate, and to a lesser extent (falling to about 65 mmHg over 90 minutes). These results are consistent with fiber reconversion occurring within the soleus muscles. The differential response is consistent with an increase in the ability of the soleus muscle fibers regaining their fatigue resistance, and correspondingly, their ability to produce the sustained contractions required to ensure adequate venous return to the heart while seated.

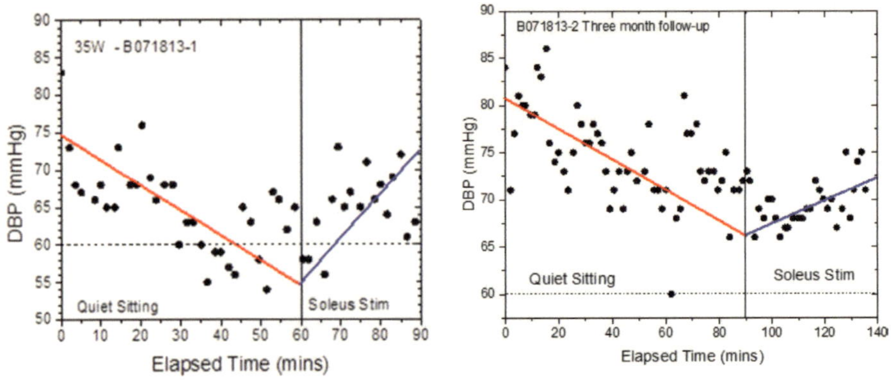

Figure 9.
Soleus muscle retraining following three months of daily, external stimulation. Left panel - In a young adult (35 y.o.) woman, sitting provides insufficient soleus muscle stimulation to sustain the venous return necessary to prevent diastolic blood pressure from falling into a severe hypotensive range. However, external stimulation of the soleus muscles is able to return diastolic pressure back to the normal range. Right panel – Following three months of daily use of soleus muscle stimulation the soleus muscles are capable of preventing severe hypotension even over a sitting duration of 90 minutes, but still unable to sustain a normal diastolic pressure.

8. Influence of soleus muscle stimulation on cognitive function

The ability of soleus muscle stimulation to normalize cardiac output and blood pressure, raises the obvious question of the extent to which such improvements in cardiovascular function can influence cognitive function. Two small pilot studies we have undertaken lead us to believe there is substantial potential for this simple, non-invasive, intervention to slow, and even reverse, the cognitive decline associated with chronic exposure to low cardiac output and the corresponding low cerebral perfusion.

In a three month study on individuals (average age of 82 years) residing in an assisted living center [23], cognitive performance was tracked weekly using the Incongruent Stroop Executive Function Test [24]. Five control subjects with normal blood pressure (resting diastolic blood pressure above 70 mmHg) and five intervention subjects with below normal resting diastolic pressure were recruited into the study. Intervention subjects self-treated to one hour per day of soleus muscle stimulation using a HeartPartner soleus muscle passive exercise device (Sonostics, Inc.). While at the start of the study, the intervention group required almost twice as long to complete the executive function test. Over the three month duration of the study, blood pressures and test times for the control group remained steady. However, the intervention group experienced improvements in both their resting diastolic pressures and their ability to complete the Stroop executive function test, such that at the end of study, test execution times matched that of the control group (**Figure 10**).

Because there is the potential for learning curve effects to play a role in traditional executive function tests such as the Incongruent Stroop when they are given repeatedly to the same study subjects over short separation times, we have also observed the influence of soleus muscle stimulation on cognitive function as assessed by a computer aided assessment which has been shown to have high repeatability and low learning curve effects, and which involves motor, memory, and executive function skills (Cognivue, Inc.). Six subjects, over the age of 65 years, who tested in the moderate to severe cognitive impairment range using the Cognivue assessment, were recruited. Each subject was provided with a soleus muscle stimulation device and encouraged to use the device for at least 2–3 hours per day. Subjects were tracked approximately every month, for six months, or until they cognitive performance returned to the normal range (Cognition score > 75).

All six subjects experienced a return to normal function during the course of the study, though the rate of return was dependent on age of the subjects (**Figure 11**). Subjects in their 60s demonstrated cognitive improvement rates of over 10%/week, while those in their 80s demonstrated improvement rates in the range of only 1–2% per week. Nonetheless, extrapolating over time, even these low rateswould mean

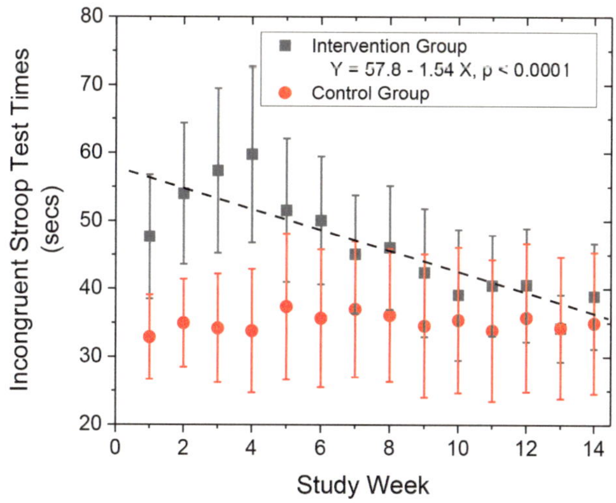

Figure 10.
Long term effects of daily soleus muscle stimulation on cognitive performance in an elderly (average age of 82) population residing in an assisted living center. Cognitive assessment relied on the incongruent Stroop executive function test. Control (normotensive) group test completion times did not vary significantly over three months. The intervention group (DBP < 70 mmHg at start of the study) received one hour per day of soleus muscle stimulation. While test completion times for the intervention group were initially almost twice that of the control group, over three months of daily soleus stimulation test times recovered to a level similar to that of the control group.

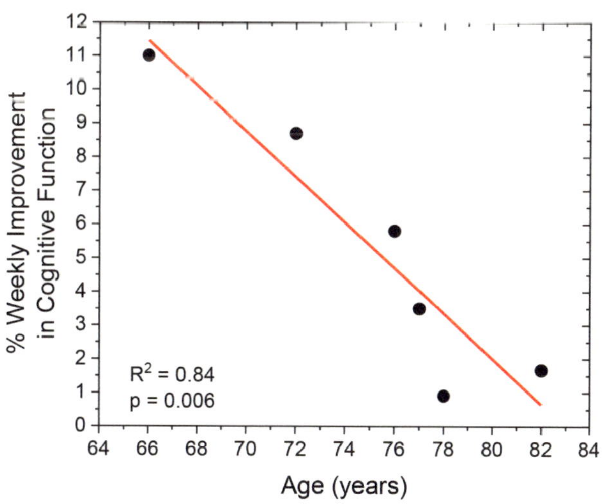

Figure 11.
Cognitive recovery rates as a function of age. Subjects with cognitive performance in the moderate to severe cognitive performance range undertook soleus muscle stimulation for 2–3 hours per day until cognitive performance reached a normal level (cognition score > 75). Individuals in their 60s experienced cognitive performance improvement at a relatively remarkable rate of 10% per week. Individuals in their 8th or 9th decades experience cognitive improvement, but at much lower rates (1–2%/week). A 2%/week cognitive improvement rate indicates that approximately 6 months of intervention would be required to move an individual from the moderate cognitive impairment level to the normal cognitive function level.

that an older individual starting out with severe cognitive impairment (Cognition score < 50) would still be able to return to normal cognitive function within a one year period of time.

9. Conclusions

The impact of dementia on both the healthcare system and society is already large and has the potential to become overwhelming in the near future. Alzheimer's Disease is the most prevalent form of dementia and the strong association between beta-amyloid accumulation in the brain and Alzheimer's provided some hope that if beta-amyloid production could be slowed, or its removal accelerated, dementia could be cured. To date, this strategy has failed to develop, and it is unclear if this strategy will be successful anytime in the near future.

As a result, the current consensus is that we need to identify a means of preventing the development of the cognitive aging which commonly progresses to dementia. Because this will require that any intervention will need to be implemented before there are indications of significant cognitive decline, successful interventions will have to be simple, inexpensive, non-invasive, and well accepted by older adults. Compliance is always challenging for healthcare interventions when the health condition is symptomless, and so it is always beneficial if the intervention produces benefits beyond the primary goal.

What has become clear over the past three decades is that reduced cardiac output, leading to reduced cerebral perfusion, is a robust predictor of cognitive aging and all cause dementia. Though cardiac output commonly declines with age, declining cardiac cardiovascular performance is not, per se, an age dependent outcome, but rather is a function of venous return. Venous return, correspondingly, is primarily dependent on the ability to maintain sufficient soleus muscle pump function whenever a person is in upright posture. The key, therefore, to maintaining cardiac output over a lifetime, is to maintain soleus function over an individual's lifetime.

Soleus muscles lose their ability to maintain adequate venous return, in large part, due to modern society's transition to chair sitting as the normal upright resting mode. Fortunately, like all muscles, the soleus muscles can be retrained and preliminary studies utilizing non-invasive soleus muscle stimulation technology has demonstrated that the improved cardiac output and normalization of blood pressure which results from soleus retraining leads to a reversal of cognitive decline even for those in their 9th decade of life. These preliminary results indicate that simple, well accepted, intervention techniques for the prevention, and even reversal, of cognitive aging, are a viable option for eliminating the devastating economic and social consequences of dementia.

Acknowledgements

The author would like to acknowledge the considerable assistance of Mr. Kyle Washington, and Ms. Linda Robertson, in the collection of data presented in this manuscript.

Conflict of interest

Dr. McLeod holds an equity position in Sonostics, Inc.

Author details

Kenneth J. McLeod
Clinical Science and Engineering Research Laboratory, Department of Systems Science, Watson College of Engineering and Applied Science, Binghamton University, Binghamton, NY, United States

*Address all correspondence to: kmcleod@binghamton.edu

References

[1] Plassman BL, Langa KM, Fisher GG, et al. Prevalence of dementia in the United States: The aging, demographics and memory study. Neuro-epidemiology. 2007;29:125-132. DOI 10.1159/000109998

[2] Kametani F, Hasegawa M. Reconsideration of amyloid hypothesis and Tau hypothesis in Alzheimer's Disease. Frontiers in Neuroscience. 2018;12:1-11. DOI 10.3380/fnins.2018.00025

[3] Sturmberg JP, Picard M, Aron DC, et al. Health and disease: Emergent states resulting from adaptive social and biological network interactions. Frontiers in Medicine. 2019;6:1-14. DOI 10.2289/fmed.2019.00059

[4] Hurd MD, Martorell, P, Delavande A, Mullen KJ, Langa KM. Monetary costs of dementia in the United States. N Engl J Med. 2013;368:1326-1334. DOI 10.1056/NEJMsa1204629

[5] Hughes ML, Agrigoraei S, Jeon M, Bruzzese M, Lachman ME. Change in cognitive performance from midlife into old age: Findings from the Midlife in the United States (MIDUS) study. J Int Neuropsychol Soc. 2018;24:805-820. DOI 10.1017/s1355617718000425

[6] Murman DL. The impact of age on cognition. Semin Hear. 2015;36:111-121. DOI 10.1055/s-0035-1555115

[7] Middlemiss JE, Cocks A, Paapstel K, et al. Evaluation of inert gas rebreathing for determination of cardiac output: Influence of age, gender and body size. Hypertension Research. 2019; 42:834-844. DOI 10.1038/s41440-018-0179-1

[8] Ricci F, DeCaterina R, Fedorowski A. Orthostatic hypotension: Epidemiology, prognosis, and treatment. J Am Coll Cardiol. 2015;66:848-860. DOI 10.1016/j.jacc.2015.06.1084

[9] Torabi P, Ricci, F Hamrefors V, Sutton R, Fedorowski A. Classical and delayed orthostatic hypotension in patients with unexplained syncope and severe orthostatic intolerance. Frontiers in Cardiovascular Medicine. 2020;7:1-8. DOI 10.3389/fcvm.2020.00021

[10] Waldstein SR, Giggery PP, Thayer JF, Zonderman AB. Nonlinear relations of blood pressure to cognitive function: The Baltimore longitudinal study of aging. Hypertension. 2005; 45:374-379. DOI 10.1161.91.HYP.0000156744.44218.71

[11] Guo A, Viitanen M, Fratiglioni L, Winblad B. Low blood pressure and dentia in elderly people: The Kungsholmen project. British Medical Journal. 1996;312:805-808. DOI 10.1136.BMJ.312.7034.805

[12] Morris MC, Scherr PA, Hebert LE, Glynn RJ, Bennett DA, Evans DA. Association of incident Alzheimer's disease and blood pressure measured from 13 years before to 2 years after diagnosis in a large community study. Arch Neurol. 2001;58:1640-1646. DOI 10.1001/archner.58.10.1640

[13] Verghese J, Lipton RB, Hall DB, Kuslansky G, Katz MJ. Low blood pressure and the risk of dementia in very old individuals. Neurology. 2003;61:1667-1672. DOI 10.1212/01.wnl.0000098934.18300.be

[14] Gabin JM, Tambs K, Saltvedt I, Sund E. Holmen J. Association between blood pressure and Alzheimer disease measured up to 27 years prior to diagnosis: the HUNT study. Alzheimer's Research & Therapy. 2017;9:37-49. DOI 10.1186/s13195-017-0262-x

[15] Gudala K, Bansal D, Schifano F, Bhansali A. Diabetes mellitus and risk of dementia: A meta-analysis of prospective observational studies. J

Diabetes Invest. 2013;4:640-650. DOI 10.1111/jdi.12087,2013

[16] Kuehn BM. In Alzheimer Research, glucose metabolism moves to center stage. JAMA. 2020;323:297-299.

[17] Young DB. Venous Return. In: *Control of Cardiac Output*. NCBI Bookshelf. National Library of Medicine, National Institute of Health. 2010. Chapter 2. DOI 10.4199/C00008ED1V01Y201002SP006

[18] Ekelund U, Tarp J, Fagerland MW, et al. Joint associations of accelerometer measured physical activity and sedentary time with all-cause mortality: A harmonized meta-analysis in more than 44000 middle-aged and older individuals. Br J Sport Med. 2020;54:1499-1506. DOI 10.1136/bjsports-2020-103270

[19] Thompson LV. Skeletal muscle adaptations with age, inactivity, and therapeutic exercise. J Orthop Sports Phys Ther. 2002;32:44-57. DOI 10.251g/jospt.2002.32.2.44

[20] Widrick JJ, Knuth ST, Norenberg KM et al. Effect of a 17 day spaceflight on contractile properties of human soleus muscle fibres. J Physiol. 1999;516:915-930 DOI 10.1111/j.1469-7793.1999.0915u.x

[21] Raichlen DA, Pontzer H, Zderic TW, Harris JA, Mabulla AZP, Hamilton MT, Wood BM. Sitting, squatting, and the evolutionary biology of human inactivity. PNAS. 2020;117:7115-7121. DOI 10.1073/pnas.1911868117/–/DCSupplemental

[22] Mansoubi M, Pearson N, Clemes SA, et al. Energy expenditure during common sitting and standing tasks: Examining the 1.5 MET definition of sedentary behavior. BMC Public Health. 2015;15:516-523

[23] McLeod KJ, Stromhaug A. Reversal of cognitive impairment in a hypotensive elderly population using a passive exercise intervention. Dovepress. 2017;12:1859-1866. DOI 10.2147/CIA.S147959

[24] Hutchison LA. Balota DA, Duchek JM. The utility of Stroop Task Switching as a marker for early stage Alzheimer's Disease. Psychol Aging. 2010;25:545-559. DOI 10.1037/a0018498

Chapter 7

Lipid Rafts and Development of Alzheimer's Disease

Mario Díaz and Raquel Marin

Abstract

A wealth of evidence accumulated over the last two decades has unambiguously linked lipid rafts to neurodegenerative diseases, in particular to Alzheimer's disease (AD). These microdomains are highly dynamic membrane platforms with differentiated physicochemical and molecular properties compared to the surrounding membrane microenvironment, and are the locus for a number of central processes in neuronal physiology. Most recent evidence pinpoint to lipid rafts as main players in AD neuropathology. It is now widely accepted that lipid rafts actively participate in the processing of amyloid precursor protein to generate amyloid beta peptides, a main component of amyloid plaques. Current evidence have highlighted the existence of severe alterations in the molecular structure and functionality of lipid rafts in the frontal cortex of human brains affected by Alzheimer's disease. An exceptionally interesting observation is that lipid raft destabilization can be demonstrated even at the earliest stages of AD neuropathology. In the present review, we will first elaborate on the structure and function of these multifaceted subcellular structures and second to focus on the impact of their alterations in neuronal pathophysiology along the onset and progression of AD continuum.

Keywords: membrane microdomains, lipid rafts, membrane neurochemistry, lipid-protein interactions, lipid raft biophysics, lipid raft aging, neurodegeneration, Alzheimer's disease (AD)

1. Introduction

1.1 Lipid rafts: definition and significance

Our current view of cell membranes is far from the historical view of a being floating mixture of lipids and proteins mixed uniformly in the form of bilayers. Instead, evidence accumulated in the last three decades has revealed that membrane constituents can segregate to form discrete domains. The heterogeneous structures of membrane lipids provide them the ability to mix non-randomly in the bilayer and to form specific lipid microdomains. The best characterized class of these structural entities has been termed 'lipid rafts' which are featured by their higher contents of cholesterol and sphingolipids compared to their surroundings. Despite some controversies on a proper definition for lipid rafts, in 2006, at the Keystone Symposium on Lipid Rafts and Cell Function, it was agreed that "membrane rafts are small (10–200 nm), heterogeneous, highly dynamic, sterol, and sphingolipid-enriched domains that compartmentalize cellular processes. Small rafts can sometimes be stabilized to form larger platforms through protein–protein and protein-lipid interactions" [1].

IntechOpen

Lipid rafts have been found in most cell types, from epithelial cells to neurons, and share essential chemical and physical properties, but differ in specific components, mainly proteins, which are responsible for functional heterogeneity of cell types, populations or even developmental stages.

The importance of lipid rafts in nerve cells lays in the fact that they behave as functional platforms which participate in a number of physiological processes involved in signal transduction, such modulation of receptor activities, protein interactions in transduction cascades, and the function of ion channels, but also in dendritic and axonal protein trafficking and sorting, regulation of neurotransmitter receptors and in the exocytotic neurotransmitter release, posttranslational modifications of proteins and lipids, and in many aspects related to cell-to-cell communication, including multifaceted synaptic physiology [2–4].

The agreement exist that these microdomains are highly dynamic structures providing transient and fluid architectural scaffolding platforms, which by undergoing structural and functional changes they accomplish a variety of functions in a coordinated intracellular and extracellular context. Remarkably, current evidence demonstrate they lipid rafts may also play significant roles in different pathological conditions. Thus, lipid rafts and raft components are key players in a variety of pathological events, i.e. by facilitating conversion of prion protein (PrP^{c}) to its infectious scrapie form (PrP^{sc}) [5], by regulation of Amyloid Precursor Protein processing in Alzheimer's disease [6], by expressing binding sites for toxins internalization such cholera toxin [7] or by providing specific entry pathways for various types of viruses and budding of mature virions from infected cells [8, 9] including the HIV-1 or the SARS-CoV-2 which is driving us mad, towards an unprecedented global chaos (by providing attachment of S-protein to ACE2 and other auxiliary proteins clustered in lipid rafts) [9], amongst other pathological processes. During the last decade, investigation on lipid rafts biology has received enormous attention due to the demonstration of its involvement in neurodegenerative diseases, in particular in Alzheimer's disease, as we will discuss later.

2. Biochemical and biophysical structure of lipid rafts

Besides being enriched in cholesterol and sphingolipids, lipid rafts are also endowed with a particular lipid signature, which makes them different from other domains in the non-raft membrane plane. Such differences are illustrated in **Figure 1** for membrane raft and non-raft fractions in the gray matter of human frontal cortex. As can be observed, compared to bulk non-raft membranes, lipid rafts contain higher contents of saturated fatty acids, lower levels of mono- and polyunsaturated fatty acids, and nearly all of the cellular contents of sphingomyelin, cerebrosides and sulfatides.

According to the polar head group, sphingolipids are divided in two major phosphosphingolipids such sphingomyelin, and glycosphingolipids which includes gangliosides, cerebrosides and sulfatides. Ceramide serves as the backbone to generate sphingolipids to produce sphingomyelin or more complex glycosphingolipids after incorporation of phosphocholine or sugars at the hydroxyl group. Both classes of sphingolipids, phosphosphingolipids and glycosphingolipids, are major components of lipid rafts and display pleiotropic behaviors affecting a number of essential functions associated with normal and pathological states, particularly in AD [10–12].

Gangliosides are acidic glycosphingolipids containing one or more sialic acid residues, representing about 6% of total lipids, and particularly abundant in raft-like lipid microdomains of neuronal cells [12, 13]. They are concentrated in the outer leaflet of the plasma membrane (**Figure 2**), where they are anchored by the

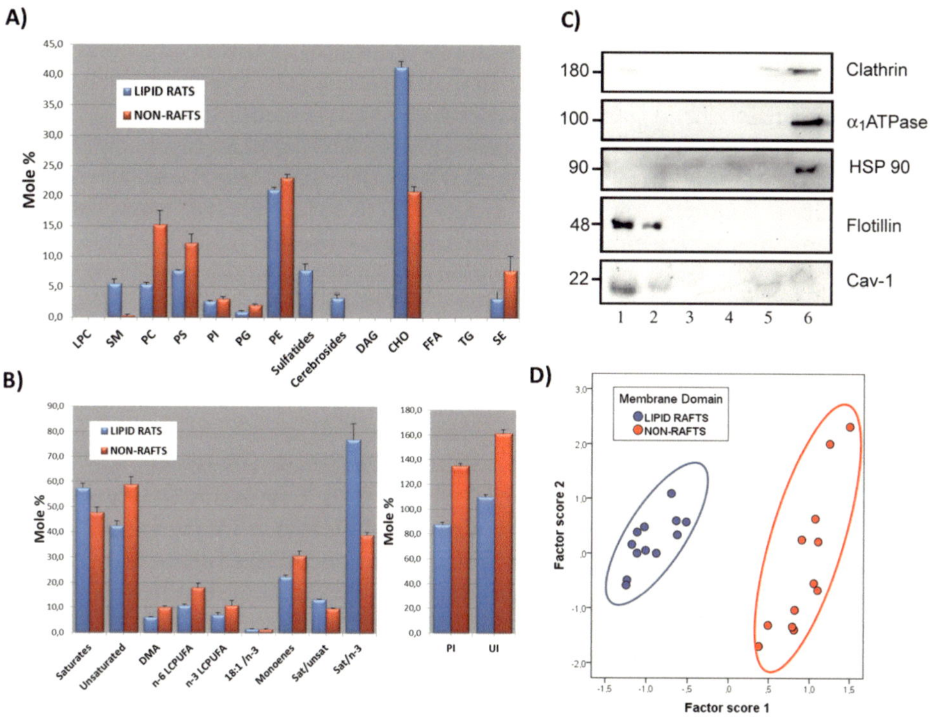

Figure 1.
Comparative analyses of lipid rafts and non-raft domain in the gray matter of human frontal cortex. (A) Lipid classes. LPC: Lysophosphatidylcholine, SM: Sphingomyelin, PC: Phosphatidylcholine, PS: Phosphatidylserine, PI: Phosphatidylinositol, PG: Phosphatidylglycerol, PE: Phosphatidylethanolamine, DAG: Diacylgycerides, CHO: Cholesterol, FFA: Free fatty acids, TG: Triacylglycerides, SE: Sterol esters. (B) Main fatty acid groups. DMA: Dimethylacetals, LCPUFA: Long-chain polyunsaturated fatty acids, PI: Peroxidability index, UI: Unsaturation index. (C) Distribution of protein markers in fractional separation of cell membranes. Fractions 1 and 2 correspond to lipid rafts while fraction 6 is mainly composed by non-raft membranes. (D) Score plots of lipids (fatty acids) resulting from multivariate analyses on membrane domains from mouse cortex illustrating the lipid fingerprints of raft and non-raft domains.

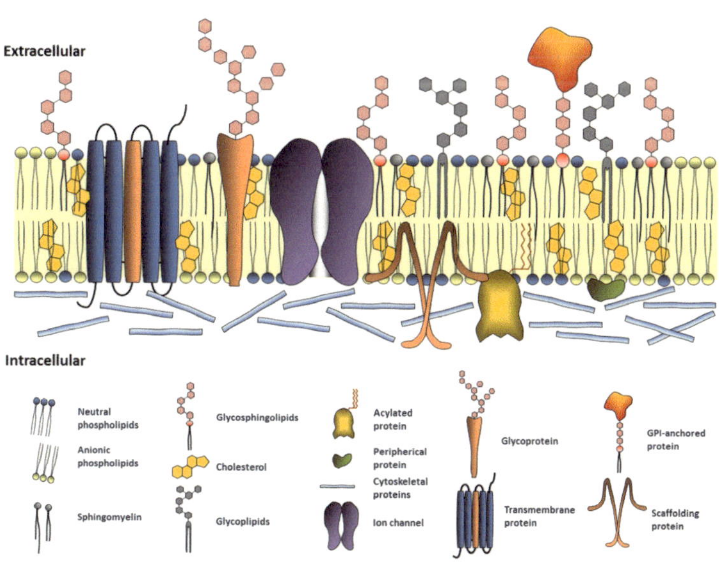

Figure 2.
Schematic representation of the lipid raft membrane structure. For details see the text.

hydrophobic ceramide part of their molecule while the oligosaccharide chain protrudes into the extracellular medium. Main gangliosides in the brain are GM1 and GDs (GD1a in a-series, and GD1b and GT1b in b-series) [11]. They participate in the two-dimensional and transverse structuration of the membrane, lipid–protein interactions and organization of lipid rafts [13, 14]. The high heterogeneity of oligosaccharide structures in gangliosides allows specific interactions with a diversity of molecules at the surface of cell membrane [10, 13, 14]. "Cis" and "trans" interactions of gangliosides play multiple roles in infectious diseases [15] where they act as cellular receptors and coreceptors for viruses, bacteria, and microbial toxins. Prominent examples are GM1 as the receptor for *Vibrio cholerae* toxin (cholera toxin), for *Clostridium botulinum* toxin (botulinum toxin), and for the SabA adhesin of *Helicobacter pylori* [15, 16]. Further, as we will discuss later, gangliosides are important regulators of amyloid β toxicity in Alzheimer's disease by modulation of polymerization of peptide species [17].

Lipid rafts biogenesis occurs in the trans-Golgi network, where their composition is set and the resulting vesicles fused to the plasma membrane. One remarkable characteristic lipid profile of lipid rafts in nerve cells is that, with the exception of gangliosides, sulfatides and sphingolipids, most lipid classes and fatty acids are present in significant amounts in raft and non-raft domains, but they differ substantially in their relative contents (**Figure 1A**). Triglycerides and free fatty acids are totally excluded from either domain. As a general rule, sphingolipids are more abundant in lipid rafts fraction and glycerophospholipids (or phospholipids) are more abundant in non-rafts fractions (**Figure 1A**). Amongst phospholipids, neutral phospholipids, phosphatidylethanolamine (PE) is the most abundant phospholipids in lipid rafts, while phosphatidylcholine (PC) and anionic phospholipids phosphatidylserine (PS) and phosphatidylinositol (PI) are present in significant proportion though less abundant than in non-raft domains. The presence of anionic phospholipids is paramount for neuronal physiology as they serve as sources for intracellular messengers and bioactive lipid mediators, i.e. inositol phosphates, diacylglycerol, eicosanoids (such prostaglandins and leukotrienes) and docosanoids (such Neuroprotectin D1) [17–22]. Fatty acids are also heterogeneously distributed between rafts and non-rafts, with saturates containing acyl chains of 16 or more carbon atoms being particularly abundant in lipid rafts. Monounsaturated (monoenes) and polyunsaturated fatty acids of the n-3 (mainly docosahexaenoic acid, DHA) and n-6 (mainly arachidonic acid, AA) series, are present at significantly lower amounts compared to non-rafts (**Figure 1B**).

DHA and AA are essential components of nerve cells membranes which esterify the sn-2 position of glycerophospholipids (mainly phosphatidylethanolamine and phosphatidylserine, the most abundant phospholipids in nerve cells). In cerebral gray and white matter, phosphoglyceride classes PE, PC, PS, PG and PI have distinctive LCPUFA profiles. Thus, ARA greatly exceeds DHA in phosphatidylinositol, whereas DHA exceeds ARA in phosphatidylserine. Brain phospholipids also include plasmalogens, which contain a vinyl-ether and an ester bond at the sn-1 and sn-2 positions, respectively [23]. As with conventional phospholipids, plasmalogens are classified according to their head group in the sn-3 position, the most abundant plasmalogens being plasmenyl-ethanolamine (PlsEtn) and plasmenyl-choline (PlsCho). The sn-2 position of PlsEtn and PlsCho display preferential esterification by LCPUFAs, a fact whereby plasmalogens are considered important LCPUFAs reservoirs of in nerve membranes.

Overall, differences in the lipid fingerprint of raft and non-raft domains (**Figure 1A** and **B**) are sufficiently different as to allow the complete discrimination of membrane domains based on a multivariate approach (**Figure 1D**). It turns out that saturated long acyl chains of phospholipids and especially sphingolipids, allow tight intermolecular packing through hydrophobic interactions within the bilayer, providing differentiated lipid complexes in juxtaposition with kinked unsaturated acyl chains of bulk membrane phospholipids. Raft lipids are held together by relatively weak non-covalent bonds, establishing a dynamic equilibrium of raft and non-raft regions in the plasma membrane. In the case of sphingolipids, these molecules interact laterally through van der Waals interactions and hydrogen bonds between their sphingosine backbones. Further, as the majority of sphingolipids contain long saturated acyl chains, their tighter intramembrane packing with associated lipids allows the formation of stable gel–liquid phase which lead to laterally segregation of sphingolipid-rich domains from their glycerophospholipid-rich surroundings [24, 25]. Structurally, this degree of lateral association is further increased by the incorporation of cholesterol, whose planar sterol ring interacts with the saturated acyl chains [26].

One key difference between phospholipids and sphingolipids is the length and saturation of their acyl chains. These acyl chains are always saturated and longer in sphingolipids than in phospholipids and allow hydrophobic interactions between the two leaflets of the bilayer [27, 28]. These molecular attributes are directly implicated not only in the formation of domains enriched in sphingolipids, but also in the coupling between the two leaflets in the rafts by interdigitation of the very long chain fatty acid between exoplasmic and cytoplasmic leaflets (**Figure 2**) and by augmenting hydrogen bonding in sphingolipid-sterol rich domains [28, 29]. This particularity is very important because it implies that lipid rafts exist as stable bilayer structures [28]. Hydrophilic interactions between phospholipid head groups are also critical as they provide physical forces for raft stability and formation of lipid shelves.

In physical terms, the more dense islands of sphyngolipid-, cholesterol- and saturated-rich domains, representing lipid rafts, exist in a liquid-ordered state ('*lo*' phase). It is widely accepted that rafts exist in nerve cell membranes in the liquid-ordered phase display limited lateral and rotational mobility in the bilayer. Cholesterol molecules intercalate filling gaps in sphingolipid packing, and increases the rigidity and molecular density of bilayers in lipid rafts due to its ability to tightly pack with saturated lipids when the *lo* phase is formed [30, 31]. The surrounding phospholipid bilayer enriched in unsaturated acyl chains exist in a state termed 'liquid-crystalline' or 'liquid-disordered, *ld*', and represent non-raft domains, in which the lipid acyl chains are fluid and disordered, exhibit much higher intermolecular mobility. The degree of disorder within this *ld* phase in nerve cells is considerable, as they contain the largest amount of polyunsaturated fatty acids (n-3 ad n-6 series) in the whole organism. Phospholipid-rich sphingolipid-poor liquid-crystalline domains (non-rafts) and sphingolipid- and cholesterol-rich liquid-ordered phase domains (rafts) exist in dynamic equilibrium in biological membranes [31, 32].

The fact that lipid rafts are in an ordered *lo* phase provides them an extremely useful property for technical purposes: they are resistant to solubilization in the cold by nonionic detergents (such as Triton X-100) and therefore can be isolated by differential ultracentrifugation as 'detergent-resistant membranes' or DRM (referring to the physical structure isolated by detergent insolubility, while the term 'raft' refer to the microdomain in the intact membranes). This has allowed the identification of proteins and lipids which display preferential (or exclusive) partitioning into rafts [32] (**Figure 1C**).

A number of proteins have been found associated to DRM and the list of candidates is steadily growing [33]. The term 'raftophilic' has been coined to refer to the preferential location of these proteins in DRM or lipid rafts (**Figure 1C**). Recently, a database (RaftProt), containing more than 47,000 entries (V2.0, 2020 version) of putative raftophilic proteins identified in mass spectrometry studies of isolated DRMs has been published [34]. Many of these proteins are not prototypical transmembrane proteins but display post-translational modifications aimed at favoring their targeting to lipid rafts (**Figure 2**). The first family of proteins described is GPI-anchored proteins. This family of proteins is anchored to the outer leaflet of the membrane through covalent attachment to a special glycolipid, glycosyl phosphatidylinositol (GPI) [35]. Amongst the GPI-anchored proteins involved in neuronal physiology, one of the best characterized is the cellular prion protein (PrP^c) [36]. PrP^c is constitutively expressed in neurons and preferentially localized in lipid rafts. PrPc is known to play different physiological roles in nerve cells, including regulation of ion channels and neurotransmitter receptors at the pre- and postsynaptic levels [37] and has been linked to the pathogenesis of prion disease as mentioned before [36]. Prion disease is characterized by the conformational modification of normal PrP^c into a misfolded and aggregated abnormal conformer, the pathogenic infectious form PrP^{sc}, [38]. Current evidence indicates that conversion into PrP^{sc} is entirely dependent on the lipid raft microenvironment [38].

Other raft-associated proteins are linked to saturated acyl chains through biochemical processes grouped as lipidation [39] (**Figure 2**). Lipidation is particularly important for membrane binding of peripheral membrane proteins (though it may also occur in transmembrane proteins). Often these proteins are directly acylated in specific residues with two or more palmitate chains, or a palmitate and a myristate chain. These lipid modifications, named S-palmitoylation and N-myristoylation, are finely regulated and determine not only the fate of modified proteins to target lipid rafts, but also contribute to their stabilization within the domain and modulate protein interactions occurring within rafts. Such post-translational modifications are commonly found in Src family of tyrosine kinases (STKs) [40] and scaffolding proteins [41]. Prenylation of proteins is also a lipidation mode for membrane association, consisting on a covalent attachment of an isoprenoid chain (either farnesyl- or geranyl-) to the C-terminus of proteins favoring their membrane association. This type of modification are common between members of the small G-proteins family, including Ras and Rab proteins involved in cellular signaling and oncogenicity [39, 42].

Common hallmark proteins of lipid rafts are caveolins and flotillins (**Figure 2**). These raft-resident proteins act as scaffolding structures within these microdomains [43, 44]. It should be mention that caveolin family was first known for its participation in the formation of caveolae, membrane invaginations involved in endocytosis and signaling commonly observed in non-neuronal cell types such endothelial or epithelial cells [45]. Soon after, caveolin-1 was shown to display high affinity for rafts, and to be consistently extracted in DRMs. Though these two families of scaffold proteins are not transmembrane proteins, they undergo palmitoylation, allowing their anchoring to the cytoplasmic leaftet of the bilayer, and have the intrinsic capacity to form lipid shells around themselves [43]. Most evidence suggests that the lipid-modified nature of these scaffolds proteins integrated in lipid rafts serve not only to aid targeting them to these domains but also to stabilize rafts themselves. In line with this, caveolin-1 tends to form high-molecular-weight oligomers which associate with each other in the plane of the membrane [44]. Further, numerous studies have concluded that these scaffold proteins help to compartmentalize specific signaling molecules within lipid rafts, and to modulate the specificity of protein interactions, with the final prospect of rapidly and selectively modulating cell signaling events [28, 46]. In this sense, the presence of caveolin-binding motifs in many raft proteins allow them to bind to the scaffolding domain of caveolin, which serve as a molecular filter to gather related signaling proteins close to each other, and to support additional protein–protein interactions [47]. In the case of flotillins, an evolutionarily conserved domain named "prohibitin homology domain" (PHB) determines the affinity for flotillins and the raftophilic nature of proteins carrying it [48]. These properties are crucial for the formation of dynamic multimolecular platforms termed signalosomes, with complex functions in normal and pathological nerve cells.

3. Lipid rafts in neuronal cell signaling

Current evidence demonstrate that neuronal lipid rafts serve as docking platforms that bring together a number of specific proteins which determine the specificity of neuronal functioning and communication. They include different families of proteins with functions as receptors, ion channels, transporters, membrane-bound enzymes, signaling proteins, interacting proteins, molecular adaptors, amongst others. They all share a special ability to interact with surrounding lipids mainly through lipid modifications, such lipidation with lipophilic anchors (*S*-palmitoylation and *N*-myristoylation, prenylation, GPI-anchoring) or to cholesterol itself or by specific domains in their secondary structure to facilitate their integration in lipid rafts, such cholesterol recognition amino acid consensus (CRAC motifs) [49] and phospholipid binding sites [50]. In general, the integration of a protein in the raft membrane initiates interaction with surrounding proteins within multimolecular complexes or signalosomes which are dynamically recruited to lipid rafts in response to specific stimuli. They are believed to rearrange into large, stable membrane rafts, and to associate to downstream signaling molecules when bound to cognate ligands, activating signalosomes, and eventually triggering specific biochemical events involved in the many facets of neuronal physiology [2–4, 32].

One of the best studied multimolecular complexes in neurons is membrane rafts in postsynaptic neurons, which along with PSDs (postsynaptic densities), are considered major sites of synaptic signaling [3, 51]. In depth proteomics analyses performed by [52] in PSD have allowed identification of a number of proteins (>150) in PSD-included lipid rafts which are exclusive for postsynaptic membrane rafts, and not shared by non-raft portions of PSD. Most of these proteins could be classified as typical raft proteins (i.e. flotillin-1 and 2, PrP^c), cell adhesion molecules (i.e. contactin, cadherin), ion channels (i.e. voltage-dependent calcium channels, inwardly rectifying potassium channels, NGF-gated Ca^{2+} channels), transporters (i.e. facilitated glucose transporter, high affinity glutamate transporter, GABA transporter protein, H^+-ATPase), kinases/phosphodiesterases (i.e. Ca^{2+}/calmodulin-dependent protein kinase II, phosphodiesterase 2A, cGMP-stimulated) and G-proteins/small G-proteins (i.e. heterotrimeric G-protein subunits, members of RAS oncogene family) [52, 53].

The scenario emerges that, at least in PSD, the high protein density in raft membranes creates a crowded environment in which lipid–lipid packing is affected by proteins, probably with a stronger effect inside the more ordered raft-like domains [54]. Even more, it could be envisaged that this dense packing would limit intradomain mobility and thereby affecting protein interactions and conformational changes. To this author, this is one of the principal reasons to explain the evolutionary selection of significant amounts of highly unsaturated long chain fatty acids (LCPUFA) in nerve cells lipid rafts, which we have consistently found in brain raft preparations from different origins [55, 56]. Indeed, brain tissue contains the largest amount of LCPUFA in the whole body, well above the adipose tissue, and more importantly, they are contained exclusively in cell membranes. Most frequent LCPUFA in nerve cells are docosahexaenoic acid (DHA, 22:6n-3) and arachidonic acid (AA, 20:4n-6), whose acyl chains contain 6 and 4 double bonds, respectively.

The amounts of these fatty acids in neuronal membranes differ between rafts and non-rafts domains, the later containing 3-4 times more LCPUFA than raft membranes [57, 58]. Even so, the degree of polyunsaturation of lipid rafts provides a sufficient degree of fluidity in the *lo* domain of lipid rafts, and a physical mechanism to ensure lipid and protein movements, such lateral and rotational diffusional rates as well as conformational displacements, required for proper intermolecular interactions [59, 60]. The regular bends introduced by the double bonds in the acyl chain of DHA and AA limit the stiffness of the packed bilayer and confer raft proteins a degree of motion freedom enough to accomplish molecular interactions. It is worth mentioning that, on a molar base, the contribution of unsaturation to bilayer fluidity is much higher for LCPUFA than for monounsaturated fatty acids, the most important in brain membranes being oleic acid (OA, 18:1n-9). In neural lipid rafts, this effect of LCPUFA on difussional rates is amplified by the unfavorable and repulsive interactions between the high cholesterol levels and polyunsaturated phospholipids [54, 61]. Overall, the lipid scenario in nerve cell lipid rafts renders them less ordered than in similar domains from non-neural cells. Importantly, destabilization of this physicochemical property of neural lipid rafts underlie dramatic consequences in lipid raft functioning in Alzheimer's disease, as we will discuss in next sections.

Perhaps largest evidence demonstrating the significance of lipid rafts in nerve cell function is neurotrophic factor signaling. Most receptors for neurotrophic factors are receptor tyrosine kinases (RTKs) residing (or recruited to) lipid rafts which are activated upon binding to the specific trophic factor and undergo activation autophosphorylation of specific tyrosine residues. Activated RTKs are docking proteins for multimolecular complexes or signalosomes that activate downstream intracellular signaling cascades through molecular adaptors which are also raftphilic. Final effectors of RTK signaling are keys for regulation synaptic transmission, differentiation, axon guidance and cell adhesion [2, 62].

Receptor tyrosine kinases observed to reside in lipid rafts include tropomyosin-related kinase A (TrkA) receptor and the low-affinity p75 neurotrophin receptor (p75NTR), which are receptors for Nerve Growth Factor (NGF) [63], IGF-1R (insulin growth factor-1 receptor) [64], EGFR (epidermal growth factor receptor) [65] or PDGFR (platelet-derived growth factor receptor) [66, 67], amongst others. Alternatively, RTKs that are not lipid rafts resident proteins, may translocate to rafts after activation, as it was initially demonstrated for glial-derived neurotrophic factor (GDNF)-mediated activation of the Ret RTK [2].

The Src family of protein tyrosine kinases (STKs) is composed by integral raft proteins widely expressed in the CNS and are particularly abundant in neurons. Src is the best studied STK, is ubiquitous but in nerve cells are expressed as different isoforms in a neuron-specific mode. It has been reported that at least five SFK members, Src, Fyn, Lck, Yes and Lyn are ubiquitously expressed in the central nervous system (CNS). These molecular transducers interact with, and participate in signaling from RTKs through different downstream pathways (MAPK, PI3-K, PKB/Akt, FAK) required to elicit neurotrophic responses such neurite outgrowth, myelination, axon guidance, proliferation and differentiation during CNS development [68–70]. In the developed CNS, STKs are involved in a number of additional functions, as diverse as regulation of neuronal apoptosis [71] or upregulation of ionotropic NMDAR (N-methyl-D-aspartate receptor) and other ion channels [72]. It is worth recalling that NMDARs are the main type of glutamate receptors that mediate fast excitatory transmission in central synapses, and are often located in lipid rafts. By modulating NMDARs, Src gates NMDAR-dependent synaptic potentiation and plasticity, critical for processes underlying learning and memory [72]. Aberrantly regulated STKs antagonize cell survival signaling pathways and induce neuronal apoptosis. Excitotoxicity is a major cause of neuronal death in acute and chronic neurodegenerative diseases [73]. This phenomenon is initiated by overstimulation of glutamate receptors, leading to sustained intracellular calcium overload and the constitutive activation of calpains, a family of calcium-activated proteases [74]. Calpain-mediated truncation of Src triggers excitotoxic neuronal death by inactivation of downstream Akt survival signaling [75]. Further to their effects on Src, overactivated calpains also affect different kinases GSK3β and CDK5, which lead to hyperphosphorylation of tau protein [76], one neuropathological hallmark in Alzheimer's disease, as we will show in next sections.

Non-conventional trophic/survival factors involved in neuroprotection have been reported in lipid rafts, one of this factors being estrogen receptors (ER) [77]. The presence of ER in the plasma membrane of nerve cells was not without controversy because classical ERs (alpha and beta) are cytosolic proteins with no affinity for hydrophobic domains. However, it is now clear that a subpopulation of ERα is associated to the nerve cell membranes, and that they are responsible for non-conventional effects of estrogens [78]. The accepted model indicates that targeting of ERs to membranes may be achieved by palmitoylation [79] and this explains the presence of ERα in lipid rafts [80]. Recent evidence shows that activation of membrane ER trigger survival signaling pathways involving transient activation of Raf-1/MEK/MAPK cascade [78] in a synergistic crosstalk with c-Src-receptor tyrosine kinase pathways [81] are neuroprotective and prevent neuronal death in models of Alzheimer's disease [77, 78, 80].

Coherent interactions of functionally related proteins have been demonstrated in lipid rafts. The new dimension of complex physiological processes but biochemically related at a nanoscale level have led to the concept of signalosomes. A pioneering study by Chadwick and collaborators (2010) in brain cortical lipid rafts form a transgenic model of Alzheimer's disease (3xTgAD mice) have shown that synaptic and signaling networks are organized into multiprotein complexes in lipid rafts, enabling coherent clustering of synergistic signaling proteins. Remarkably, significant alterations in numerous receptor/cell signaling protein associations were detected in the transgenic AD model [82]. These finding are quite relevant for the disease in humans, as synaptic dysfunction is one of the hallmarks of AD [3, 83, 84].

Similar signaling platforms operate in neuronal mechanisms involved in neuroprotection against different toxic insults (including amyloid β). The signalosome described by our group in lipid rafts from human frontal cortex is particularly relevant because its implications in neuronal survival and death. Our recent research indicates that lipid rafts are the site of formation of a complex set of interactions between survival/growth factors ERα (described above) and IGF-1R, scaffolding Cav-1, NMDA receptor regulator PrP^c (physiological role), and ion channels pl-VDAC (a plasmalemmal form of mitochondrial voltage gated anion channel VDAC1) and NMDAR. This ER-signalosome likely contains signal transducers such heterotrimeric G-protein and STK such Raf-1 involved in neuronal ERα signaling. Unlike in other signalosomes, in this case, proapoptotic protein (pl-VDAC) share a common cluster with survival factors [80, 85, 86]. pl-VDAC has been found as a resident protein of lipid rafts in hippocampal and septal cell lines, mouse hippocampus and frontal cortex, and human cognitive areas, such as frontal cortex, septum and hippocampus [85, 86], suggesting that location of VDAC in neuronal rafts may be a general phenomenon. Although the exact role of this mitochondrial channel in cell membrane lipid rafts is still under debate, pl-VDAC has been claimed to participate in the extrinsic apoptotic pathway [87]. The presence of pl-VDAC (and perhaps NMDAR) in lipid raft ER-signalosome suggest that they might be a critical site involved in neuronal fate decision, a fact that might be relevant in AD neuropathology, as discussed in the next sections.

The preferential location of different neurotransmitter receptors (NTR) and ion channels has been demonstrated steadily since the discovery of lipid rafts in nerve cells, and the number of candidates keeps growing and far from being definitive. An excellent review and overview of neurotransmitter receptors and transporters associated with lipid rafts in neurons and glial cells can be found in [3]. The range of NTR involved is all-encompassing and include ionotropic receptors, such AMPA-R (α-amino-3-hydroxy-5-methyl-4-isoxazole-propionic acid receptor), NMDA-R, $GABA_AR$ (γ-aminobutyric acid receptors) or nAChR (nicotinic acetylcholine receptors), metabotropic receptors such mAChR (muscarinic acetylcholine receptors), 5-HT-R (serotonin receptors) or mGluR (glutamate receptors), neurotransmitter transporters such EAAT (excitatory amino acid transporters), and many GPCR and G-proteins such Gα and Gβγ (See [3] and references therein).

Often, these proteins are not stable raft-resident proteins but behave dynamically and may traffic into or out lipid rafts, undergoing stimuli-induced integration in lipid rafts, which allow them interacting with transducers and even effector proteins (i.e. GPCR, G-protein-coupled receptors) or multimer formation (i.e. ionotropic receptors) to trigger either increase or dampen signaling responses. Lipid rafts also participate in the formation of neurotransmitter receptor clusters (i.e. NMDAR and $GABA_AR$ in postsynaptic neurons) influencing synaptic function, and are sites for endocytosis and trafficking of NTR. During neurotransmitter signaling, many GPCRs undergo agonist-induced endocytosis, leading to receptor recycling, sequestration and downregulation through clathrin-independent mechanisms [88, 89].

A number of neurotransmitter-independent ion channels have been found either as proper residents or transiently associated to lipid rafts. These proteins are generally downstream effectors of neuronal signaling and include the large family of voltage-dependent (Kv) potassium channels, Ca^{2+}-activated K^+ channels, different subunits of voltage-dependent sodium (Na_V) and calcium channels (Ca_V), VDAC and ClC families of voltage-gated chloride channels, G-protein-gated or cyclic nucleotide-gated (CNG) ion channels, transient receptor potential (TRP) cation channels, and water (AQP) channels [3, 52, 85, 90–92].

Structurally, ion channels are diverse in their architecture and topology, but generally contain several transmembrane domains, which allow integration within lipid rafts mainly through hydrophobic lipid-protein interactions with acyl chains of surrounding lipids within the lipid bilayers, and also with phospholipid head groups at the level of intra- and extracellular aqueous interfaces, being these lipid interactions absolute requirements for proper channel gating.

Further, members of all major ion channel families have been demonstrated to be regulated by membrane cholesterol and to partition into cholesterol-rich membrane domains [93]. Stability of channel proteins in lipid rafts is ensured by hydrophobic interactions of transmembrane domains with cholesterol within the core of the bilayer. Cholesterol itself not only provide channel compartmentalization but also alters the kinetic properties and current–voltage dependence of many voltage-dependent channels, particularly in voltage-gated Na^+ and K^+ channels [93]. Targeting of ion channels is isoform-specific, as demonstrated for Kv channels Kv2.1, Kv1.4 and Kv1.3, which are present in distinct membrane compartments in hippocampal and cortical cells [90, 94]. Apparently, membrane domain cholesterol levels differentially modulate the trafficking and localization of Kv channels [90, 95]. Overall, ion channels targeting to lipid rafts channel induces not only clustering to other raftphilic partners, but also modulation of ion channel gating by virtue of microenvironmental cholesterol.

Finally, lipid rafts are associated with cytoskeleton [96]. Interactions with cytoskeletal components (actin, tubulin, vinculin, filamin, and tau) contribute to the regulation of lipid rafts assembly and clustering. The accepted view is that this depends on raft lipids, raft scaffold proteins and submembrane actin network [96, 97]. It has been proposed an intuitive model ("picket-fence") whereby actin filaments anchored to cytofacial leaflet of lipid rafts regulate lateral diffusion of adjacent membrane lipids and proteins [97]. Anchoring of actin filaments with raftphilic proteins have been proposed to allow the transient clustering and coalescing of small rafts to form larger homo - and hetero -GPI-anchored oligomeric rafts, through raft-based lipid interactions that generate functional raft domains [98]. This spatiotemporal microdomain clustering depends upon cholesterol, sphingolipids, phosphoinositol lipids and the cortical actin meshwork, where actin filaments cross-linked by myosin motors promote energy-dependent lateral movements of GPI-anchored proteins [98]. Besides actin, tubulin is also associated to lipid rafts and co-precipitates with caveolin-1 in brain extracts. In fact some authors have suggested that tubulin itself might behave as a scaffolding protein in lipid rafts [3]. Recent studies have shown that lipid rafts lipids are important elements in the interaction membrane-cytoskeleton. Thus, It has been shown that phosphoinositide lipids such as PtdIns(4,5)P2 and PtdIns(3,4)P2, which accumulate in the inner leaflet of lipid rafts, bind actin and direct actin assembly into filaments, and that PtdIns (4,5)P2 serves as a tubulin anchor on the plasma membrane [3, 96]. It is known that microtubule dynamics participate in raft-associated neurotransmitter signaling [3]. Certain G proteins have been shown to promote the GTPase activity of tubulin and to affect microtubule arrangement. The association between tubulin and heterotrimeric G proteins has been demonstrated to potentiate adrenergic and cholinergic signaling neurotransmitters by directly activating their respective G proteins [3]. Further, cytoskeletal dynamics and its interaction with lipid rafts are demonstrated to be directly involved in processes such neuronal growth, axonal/dendritic guidance, axonal regeneration and dendritic spine formation in hippocampal neurons [3, 96].

4. Lipid rafts in established AD

4.1 AD neuropathology: linking in lipid rafts disturbances

Alzheimer's disease is the most common neurodegenerative disease, and has reached pandemic proportions in developed countries. AD is characterized by progressive memory loss, cognitive deficits and subsequent gradual but relentless dementia. In the majority of cases AD occurs late in life and without a known cause (referred to as "sporadic" or "late-onset" Alzheimer's disease, LOAD). Brains of individuals with AD exhibit massive loss of synapses and neurons, as well as extracellular senile plaques (SPs) and intracellular neurofibrillary tangles (NFTs). The most severe neuropathological changes occur in the hippocampus, followed by the association cortices and subcortical structures [99, 100].

The major proteinaceous component of SPs is a 40–42 amino acid polypeptide amyloid-β derived by proteolytic cleavage from the transmembrane protein APP. According to the amyloid cascade hypothesis (first proposed by Hardy and Higgins, 1992) [101], in the amyloidogenic pathway, the β-secretase activity of BACE1 (β-site APP cleavage enzyme 1), generates the amino terminus of Aβ [102] while γ-secretase complex (made up of four proteins: presenilin, APH-1, PEN2 and nicastrin) [103] cleavage at the carboxy-terminus and determines its length (Aβ40 and Aβ42). Aβ40 is the most common species and Aβ42 the more fibrillogenic and prone to aggregates in SPs [104]. As discussed below, lipid rafts are the key membrane domain for this sequential cleavage of APP. Conversely, in the non-amyloidogenic pathway, the zinc metalloproteinase ADAM10 named α-secretase, cleaves APP within the Aβ domain [105, 106] and thus precludes Aβ formation. Of note, action of α-secretase occurs predominantly in non-raft domains [105, 106]. A reciprocal relationship exists between non-amyloidogenic and amyloidogenic APP processing such that impaired ADAM10-mediated proteolysis of APP serves to enhance amyloidogenic processing thereby elevating levels of Aβ peptides in AD-afflicted brains [105]. These assertions are extremely important for AD onset and progression, since it suggests a dynamic intramembrane and interdomain competition between proteolitic activities of α- and β- secretases on APP, which largely determine the balance between amyloidogenic and non-amyloidogenic patwways. Further, increasing evidence indicates that ADAM10 may also affect AD pathology through potential mechanisms including reducing tau pathology, maintaining normal synaptic functions, and promoting homeostasis of neuronal networks [106].

Fibrillation of amyloid β peptides is a critical and complex process largely responsible for amyloid toxicity [107]. The available evidence favors a model in which the conversion of the normally soluble Aβ peptide into insoluble oligomeric, protofibrillar, and fibrillar toxic forms [104, 107].

NFTs are also a pathological hallmark of AD, though not exclusive. NFT are also present in other taupathies such frontotemporal dementia [108]. NFTs contain abnormally hyperphosphorylated forms of the microtubule-associated protein tau, which causes detachment of tau from microtubules and the formation of insoluble tau aggregates. This leads to the occurrence of paired helical filaments and NFTs present in cell bodies and apical dendrites as neurofibrillary tangles (NFTs), but also in distal dendrites as neuropil threads, or in abnormal neurites associated to SPs [109].

A number of studies demonstrate that AD-associated cognitive dysfunction is strongly correlated with the accumulation of amyloid-β and hyperphosphorylated tau, however, the precise relationship(s) between neurological and biochemical hallmarks of AD remains incompletely understood, particularly in Sporadic Alzheimer's disease. Likewise, potential causes for AD remain unknown, except for the familial form of the disease (familial AD, FAD), in which several genetic mutations on proteins involved in amyloid β production have been well-identified [84, 110]. In FAD, which accounts for less than 1% of the total AD cases, rare autosomal dominant mutations have been identified in three genes, namely APP, PSEN1 and PSEN2, the latter two being the most common mutations found in FAD [84, 110].

Regarding the relationship between alterations amyloid-β and tau in AD-afflicted brains, current models suggest that amyloid oligomerization and aggregation drives tau hyperphosphorylation and fibrillation. By itself this modified form of tau stimulates cell dysfunction and neurodegeneration in AD both downstream and independently of Aβ [84, 111].

4.2 Lipid rafts and amyloid processing

Numerous studies have shown that proteins involved in amyloidogenic processing pathway APP, BACE1 and the γ-secretase complex are transmembrane proteins associated to different extents to lipid rafts (see the excellent review by Hicks et al., 2012) [6]. APP is localized in raft and non-raft fractions, but predominates outside rafts, while the β- and γ-secretases are mainly located in rafts. Noteworthy, APP and secretases exist in two pools, raft and non-raft, but their relative residence fraction vary depending on cellular signals and physicochemical microenvironmental factors (i.e. lipid composition of bilayer, see below). Conversely, the α-secretase involved in non-amyloidogenic pathway is membrane- but not raft-associated [106].

Regulation of APP raft localization involves interaction between the C-terminus of APP and flotillin-1 [112]. Further, another factor promoting raft localization of

APP is cholesterol. APP specifically binds cholesterol though a direct interaction with the C-terminal domain C99 (also known as β-C-terminal fragment, β-CTF) [113]. Apparently, binding of cholesterol to C99 would favor the amyloidogenic pathway in cells by promoting localization of C99 in lipid rafts [113].

Regarding secretases, raft localization of β-secretase and interaction with raft-resident lipids is mediated by palmitoylation of BACE-1 [114]. It has been reported that when BACE-1 is targeted to lipid rafts via GPI-anchoring, upregulation of amyloidogenic APP processing occurs and production of Aβ is increased [115]. Subunits of the γ-secretase complex are enriched in lipid rafts by means of S-palmitoylation of nicastrin and Aph-1 [116, 117] but, interestingly, does not directly modulate γ-secretase processing of APP [117]. Further, the lipid raft scaffolding protein caveolin-1 influences the γ-secretase spatial distribution favoring its partitioning to lipid rafts but also enhances its secretase activity [118].

Another important lipid-raft associated protein which was shown to play an important role in APP processing is PrP^c. Indeed, PrP^c regulates APP processing by inhibiting β-secretase activity in the cell surface, and this effect requires the localization of PrPc to lipid rafts [6, 119]. This led to the hypothesis that PrPc might be a key protective protein against AD, and that PrPc downregulation might impede the negative control of BACE1 activity and accumulation of Aβ peptide. However, no decrease of PrPc content has been reported in AD brains, therefore it is suggested that decreased ability of PrPc to control BACE1 might be consequence of age- and disease-dependent disruption of lipid rafts, at least in the case of sporadic AD [6].

The relationship of amyloid peptides and membrane components is often reciprocal. Several examples illustrate this bidirectional relationship. First, PrPc has been shown to be a receptor for Aβ oligomers (even at nanomolar concentrations). Binding of Aβ oligomers to PrPc results in the blockage of hippocampal LTP and reduction of PrP affinity for the NMDAR (through a complex allosteric modulation of its glycine binding site). Once out of the control by PrP, this results in steady-state NMDAR currents and excitotoxicity [6, 120]. Together with BACE1 regulation by PrPc explained above, this provides an integrated toxicity mechanism explaining the interplay between BACE1, PrPc, NMDAR, Aβ species and hippocampal LTP, in the hippocampal degeneration and functional decline in AD.

A second example is brain cholesterol. Within nerve cells, the biggest reservoirs of cholesterol are found at the plasma membrane, myelin sheaths and in the endocytic recycling membranes. The majority of brain cholesterol is derived from de novo biosynthesis, rather than from plasma LDL [121]. Cholesterol can directly modulate amyloidogenic secretase activities leading to altered amyloid-β generation [10, 122–124]. Collectively, these data indicates that elevated cholesterol levels promote the co-clustering of APP and BACE1 in lipid raft domains, as well as their rapid endocytosis, and increases their activities. Conversely, experimental reduction of membrane cholesterol levels decreases the association of BACE1 with lipid rafts and reduces the activity of both BACE1 and γ-secretase, leading to additive reduction of amyloid-β production.

Cholesterol levels in the brain are regulated through a series of steps in a cross-talk between astrocytes and neurons (see excellent reviews by [125–127]). These involve HMG-CoA reductase (HMG-CoA, the rate-limiting enzyme responsible for cholesterol synthesis in neurons and glial cells), APOE-containing HDL-like particles released from astrocytes (which mediates the uptake of lipoprotein particles via LRP), LDL receptor-related protein (LRP, which serves as a neuronal receptor for astrocyte-produced APOE-containing lipid particles), ATP-binding cassette sub-family A member 1 (ABCA1, mediating cholesterol efflux from neurons has been also shown to modulate Aβ levels in neurons), and acyl CoA:cholesterol acyltransferase 1 (ACAT1, which converts free cholesterol into cholesteryl esters), amongst other proteins. Excess free cholesterol in neurons is either converted to cholesteryl esters by ACAT1 or exported through ABCA1. Several lines of evidence indicate that cholesterol efflux, synthesis or esterification controls amyloid-β generation. Thus, stimulation of HMG-CoA or ACAT1 has been demonstrated to increase Aβ levels though mechanisms still poorly understood. Further, in vivo studies have shown that deletion of ABCA1 gene decreases the levels of APOE, a finding that correlates with greater amyloid-β deposits. Moreover, increased intracellular cholesterol (and perhaps cholesterol esters) has a considerable impact on membrane domain biogenesis and lipid raft formation, eventually leading to stimulation of amyloidogenic APP processing [128–130].

4.3 Untangling the conundrum of late-onset AD origin

Despite intense scientific research in the areas of genetics, molecular and cell biology, and neuroscientists throughout the world, causative factors for nerve cells destruction in LOAD are far from conclusive and have not been definitively established. Amongst factors evidencing solid links with neuronal loss and development of sporadic Alzheimer's disease are genetic polymorphisms, such ApoE4 [102, 103], neuroinflammation [104–106, 131], oxidative stress [107–110], neurolipid deregulation [111–114, 131], environmental factors, such chronic exposure to neurotoxic metals, pesticides or nanoparticles [115–117], dietary habits [118–121], and xenoendocrine and hormonal changes such menopause [1, 122, 123]. However, the only factor that is unequivocally associated to the onset of AD is aging. Aging is an extremely complex biological process affecting whole organism. Cerebral aging is acknowledged to involve multiple factors which converge to reduce cognitive functions such as mental speed, executive function, episodic memory, working memory, short-term recollection, spatial memory and capability to process new information, amongst other deficits [92, 124, 125]. These cognitive deficits are recognized to be secondary to losses in synaptic contacts, reduced neuroplasticity, dendritic branching, changes in neuronal and/or astrocyte physiology and crosstalk [126], and is accompanied by reductions in the volume of the hippocampus and pre-frontal, parietal, temporal and entorhinal cortical parenchyma [92]. Not surprisingly, brain areas which are more neuroplastic throughout life, such hippocampus and entorhinal cortex are most vulnerable to age and more prone to undergo pathological neurodegeneration [126]. Indeed, neurons that are particularly vulnerable in AD include the pyramidal layers of the hippocampus, those in layer II of the entorhinal cortex, and from certain areas of the neocortex (frontal, parietal and temporal cortices) [92]. Although most vulnerable neurons use glutamate as neurotransmitter (the most common in the brain), there is also significant loss cholinergic and noradrenergic neurotransmission in subcortical neurons in the basal forebrain [127]. In particular, the dysfunction of cholinergic neurons has received much attention (as per involved in obvious deficits in attention and memory in AD) and has been the stem for the "cholinergic therapy" in AD [127] Current knowledge support the notion that much of the cognitive dysfunction in AD is not due to loss of neurons containing a particular neurotransmitter, but to disruption of the network connections between key brain regions within the limbic system and specific areas of the neocortex [79].

My current view, shared with most neurologists and molecular and cellular neurobiologists, is that LOAD onset is determined by the slow but steady deleterious contribution of a combinatorial concert of factors referred above, superimposed to, and facilitated by both genetic predisposition and the exhaustion effect of lifestyle and aging. For instance, it is known that the apolipoprotein E allele e4 (*APOEε4*) expressed in the brain is a genetic risk factor for LOAD, whereas the e2 allele is protective. One copy of *APOEe4* increases the risk for AD by ~3-fold and two copies by ~12-fold (http://www.alzgene.org), but its effect is magnified by aging, with a decrease in age at onset by ~5 years/e4 allele, in both sporadic and familial forms of AD [84, 132–134]. In line with this, the Society for Women's Health Research Interdisciplinary Network on AD, comprised of an expert panel of scientists and clinicians, has reviewed ongoing and published research related to sex and gender differences in AD, and defined the concert Age-APOE-Gender a triad of high risk for AD [133].

4.4 Lipid rafts: beyond and before amyloid possessing

The involvement of lipid rafts in AD extends well beyond facilitating amyloidogenic processing of APP or tau hyperphosphorylation. As described above, numerous neurotransmitter receptors, neurotrophic factors receptors and downstream signaling proteins, signalosomes, membrane trafficking components, ion channels and pathway effectors have been demonstrated to be differentially altered in Alzheimer's disease. Indeed, the number of cellular and molecular biological processes known to be presumably affected in AD is enormous. It is conceivable that not all these evidences occur in real degenerating human brains, as most observations have been obtained under artificial in vitro conditions, or in vivo using cellular and animal models, often overexpressing human proteins not normally expressed in experimental animals. These same arguments may also explain why contradictory results or fundamental controversies from different research groups are found in the literature. Furthermore, very relevant information from studies aimed at disentangling the pathological mechanisms for AD has been obtained from transgenic mice models expressing human components of the amyloidogenic pathway from well-established mutations in familial Alzheimer's disease. Thus, even if overexpressing transgenic models may render a disease scenario to closely resemble human amyloid and/or tau pathology, results are not necessarily translatable to the most common form of AD, i.e. LOAD. One plausible hypothesis which may assemble much information on the different mechanisms reported as altered in Alzheimer's disease is that they may belong to a programme of sequential set of events triggered at the onset of the disease, in some kind of self-destructive parallel domino effects, which are exacerbated during the progression of the disease.

In this sense, plentiful and compelling evidence point to lipid rafts alterations as a common underlying factor related to AD neurodegeneration, even at very early stages of the disease. Moreover, it is now clear that these structures undergo aging-associated modifications in brain areas even in subjects without signs of the disease. Overall, this suggests that it might be disentangling of lipid rafts a very early event in the transition from normal aging to developing this neurodegenerative disease.

It may be assumed that altered function of biochemical components integrated within lipid rafts may be secondary to destabilization of membrane structure of lipid raft, in particular with neurolipids. Indeed, a considerable number of studies demonstrate that lipid biochemical and biophysical anomalies lead to abnormal functioning of lipid rafts [10, 135–137]. These issues are discussed in the next section.

4.5 Lipid abnormalities and lipid rafts dyshomeostasis in AD

In the seminal description of the degenerative disease in 1932 named after him, Alois Alzheimer highlighted the occurrence of 'adipose inclusions' or 'lipoid granules' as the third pathological hallmark of AD. This finding did not receive enough attention until recently. Subsequently, biochemical alterations of lipid composition have been reported in post-mortem brains from individuals with AD. Perhaps, the intimate link between lipid metabolism and AD was only boosted when the ε4 allele of the

APOE gene was identified as a strongest genetic risk factor for LOAD [130, 134, 138]. The involvement of lipids in AD is substantiated by a number of epidemiological studies which support a role for cholesterol and essential fatty acids in the pathogenesis of AD [138, 139]. It is now well established that most, if not all, classes of lipids are implicated in AD pathogenesis. (recently reviewed in Chew et al., 2020) [140].

A wealth of studies have consistently demonstrated the depletion of LCFUFA in brain tissue from postmortem AD brains, in particular for fatty acids of the n-3 series, mainly docosahexaenoic acid (DHA) [131–144]. As mentioned before, brain is the organ containing the largest amount of DHA in the whole organism, and its depletion, underlie many alterations occurring during AD neurodegeneration. Indeed, DHA is a pleiotropic molecule. It is an essential component of nerve cells membranes associated to glycerophospholipids (mainly phosphatidylethanolamine, the most abundant phospholipid in nerve cells), and is largely determinant of physicochemical and biophysical properties of plasma membrane, such membrane viscosity, lateral mobility, phase separation and microdomain segregation, conformational transitions and lipid-protein and protein–protein interactions [60, 145, 146]. Besides, DHA is an active modulator of neurogenesis, synaptogenesis and neurite outgrowth and in memory consolidation processes [147, 148], but also in the activation of survival signaling pathways against oxidative and proinflammatory insults, amyloid β production [149–151], and transcriptional activation of neuronal antioxidant systems [152, 153]. The importance of DHA for brain health is highlighted by the extensive epidemiological and experimental evidence linking its depletion with the development of neurodegenerative diseases [154, 155].

Another evidence linking LCPUFA and AD is that LCPFA, especially DHA and AA are highly susceptible for oxidative stress. The high metabolic rate and elevated oxygen consumption in brain tissue, together with the enrichment in redox transition metals, such iron and copper, favor the free radical-induced peroxidation of LCPUFA in the brain parenchyma [156–158], and generation of reactive lipo- / endo-peroxides such isoprostanes, neuroprotanes, malondialdehyde, acrolein, and reactive aldehydes such HHE and HNE [159, 160]. Further, unlike other forms of free radical injury, lipid peroxidation is self-propagating and generated lipoperoxides react with membrane LCPUFA to produce additional reactive lipo-endoperoxides, to provoke extensive brain tissue damage [157, 159]. Obviously, one main outcome of lipid peroxidation is the structural damage of membranes, which impairs nerve cell physiology and finally causes cell death.

Pioneering studies published by our group on lipid rafts from human frontal cortex have demonstrated altered lipid profiles in AD brains at advanced stages V-VI, compared to control brains [161]. Amongst other alterations, lipid rafts displayed abnormally low levels of n-3 long chain polyunsaturated fatty acids (LCPUFA) and unsaturation and peroxidability indexes. LCPUFA, mainly docosahexaenoic acid (DHA; 22:6n-3), are particularly enriched in nerve cell phospholipids, and their presence is an absolute requirement for neuronal membrane function [125, 146, 162]. The results in this study were relevant for two main reasons. First, lipid rafts showed that, even in non-AD subjects, neuronal lipid rafts contain significant amounts of polyunsaturations in the form of n-3 and n-6 acyl chains, which makes them less packed and ordered than supposed. These findings are not surprising as fatty acids have the capacity to influence plasma membrane organization to facilitate intermolecular mobility (in a ′crowed′ protein environment such neuronal lipid rafts) by modulating membrane lipid composition, which affects functionality of lipid raft domains [145, 162]. Second, no changes in cholesterol were associated to lipid rafts in advanced stages of AD, which apparently contradicted the observation that AD brains contained higher cholesterol levels than normal brains. However, these observations are reconcilable on the basis that bulk brain cholesterol may increase by affecting non-raft domains, without change in lipid rafts. In this case, interaction of rigid sterol ring of cholesterol with membrane phospholipids renders non-raft domains less fluid than normal, a notion which is supported by biophysical observations [59, 60, 128].

Other important rafts-associated lipids are gangliosides. These glycerospingolipids have been demonstrated to play a role as assembly- and aggregation-promoting factors [11, 17]. Aberrant levels and significant regional differences in the distribution of specific gangliosides have been observed in AD brains [10, 149]. Gangliosides are primary modulators of amyloid-β aggregation in AD, and it has been demonstrated that binding of GM1 to amyloid-β trigger conformational changes towards more ordered structures with increased β-sheet content, which correlates with higher toxicity [17, 163, 164]. A number of studies have revealed that gangliosides accumulate in senile plaques favoring the conversion of Aβ to a neurotoxic oligomers, and accelerates the formation of amyloid fibrils [152, 165, 166], these effects being favored in the presence of the ApoE4 genotype [167]. It has been demonstrated that Aβ has a high affinity for GM1 containing membranes both in vitro and in vivo, and that the N-terminal region of Aβ promote interactions with GM1 clusters in lipid rafts through hydrogen bonding and electrostatic interactions [13, 168, 169]. Further, the participation of gangliosides in the development of Alzheimer's disease is further strengthened by that fact that GM1 content in neuronal membranes, particularly in raft microdomains, increases with age [6, 152, 170]. In this sense, lipid raft GM1 acts as a 'seed' for amyloid-β aggregation [10, 151].

5. Lipid rafts alterations at early stages of AD

The presence of biochemical and physicochemical alterations in lipid rafts at early stages of AD has been recently reported [60, 171]. It is noticeable that lipid rafts are profoundly altered in the cortex of AD brains from the earliest stages namely AD I/II [172]. These changes affects the lipid matrix of lipid rafts well before the overt of clinical signs, and are retained as the disease progresses towards more advanced stages (stages III-IV) with little modifications. The most dramatic changes observed were the reductions in polyunsaturated arachidonic and docosahexaenoic acids, cholesterol, sphingomyelin, monounsaturated oleic acid, as well as increased levels of phosphatidylcholine and sterol esters [152]. Other reports have also shown elevated ceramide levels are and reduced sulfatides at the earliest clinically recognizable stage of AD [173], likely involved in oxidative stress-induced neuronal death.

Paralleling these changes, lipid rafts from AD frontal cortex displayed abnormally low unsaturation and peroxidability indexes, suggesting a high impact of lipid changes in physicochemical conditions of lipid rafts [60]. Lipid abnormalities in lipid rafts likely have a profound impact on membrane physicochemical properties, in particular to membrane order and microviscosity. We have shown that the reduction in n-3 polyunsaturated and the increase in saturated fatty acids, results in augmented density of hydrophobic interactions between saturated hydrocarbon acyl chains of phospholipids and sphingolipids within the membrane plane [11, 54, 55]. The consequences are: laterally condensed and more packed membranes, and higher physical order and microviscosity, in spite of the reduction in cholesterol [55]. These findings are in agreement with the observations in lipid rafts from the neocortex of aged APP/PS1 mice reported recently [54], which display a similar increase in membrane microviscosity secondary to reduced n-3 LCPUFA and cholesterol levels, as determined by steady-state fluorescence anisotropy [59]. Moreover, we have demonstrated that this transition towards more ordered membranes occurs during the initial stages of the pathology, and that it is correlated to the alterations observed in the lipid profiles. A finding that is retained in intermediate stages of AD. The impact of these biophysical observations are likely relevant on the dynamics of amyloid aggregation. Indeed, it is known that interaction of Aβ with neural membranes is energetically more favorable in liquid-ordered membranes than in liquid-disordered counterparts and also that this association accelerate fibrillation [119, 158, 159, 174]. The relationship between liquid ordered-membranes and amyloid peptide association is reciprocal. Indeed, studies performed in rat synaptic membranes and in human brain tissue have shown that different Aβ peptides reduce membrane fluidity by partitioning into the hydrophobic core of membranes [119, 158, 159] thus adding additional membrane order to lipid rafts.

One relevant consequence of altered physicochemical properties of lipid raft observed in human brain cortex is that these likely modify interactions between raft resident proteins, in particular those involved in the differential processing of APP (see below).

Surprisingly, anomalies in lipid rafts from early AD stages are clearly more severe than those found in late stages (V/VI) [147]. It can be speculated that the neuronal metabolic collapse and/or disruption of neuronal lipid homeostasis [175, 176] in late stages of the disease, overcome membrane biosynthetic mechanisms to maintain lipid raft structure. In turn, this would weaken the thermodynamically unfavorable boundaries and tension line between raft and non-raft domains, eventually leading to more homogeneous membranes [11, 22, 24, 28].

Noticeably, we observed that lipid rafts alterations specifically affect frontal and entorhinal cortices in the same subjects, two brain areas particularly affected in AD, while no substantial effects are observed in the cerebellum. Further, Noteworthy, alteration in neurolipid levels and biophysical properties occurs in the frontal cortex at stages I/II, a brain region that devoid of neuropathological hallmarks of AD (neurofibrillary tangles and senile plaques) at such early phase [156, 157]. Moreover, It is worth mentioning that, at least in the frontal cortex, no astroglial proliferation is present at stages I/II, and very little at stages III/IV and mainly associated to senile plaques [156, 157]. Therefore, changes in lipid composition in lipid rafts in frontal cortex at early stages of AD pathology reflect modifications in the lipid composition of lipid rafts in neurons and cannot be explained by modifications in the neuron/astroglial ratio.

We extended our lipid analyses in the frontal cortex to entorhinal cortex and cerebellum, two other brain areas differentially affected in AD [109, 177, 178]. The results showed that alterations in lipid raft found in cortex are also present, and to a similar extent and disease-course, in entorhinal cortex [152]. It is known that enthorinal cortex is one of the first brain areas affected in AD, which exhibits the neuropathological traits at stages I/II [156, 157]. Overall, the fact that frontal cortex lipid rafts exhibit altered lipid profiles at stages AD I/II but not AD neuropathological hallmarks indicates that lipid raft destabilization develops well before the appearance of neurofibrillary tangles [100, 156, 157].

We have further explored the pathophysiological consequences of these alterations in the amyloidogenic pathway during development and progression of AD. As expected, we have detected main components involved in amyloidogenic pathway, namely APP, β-secretase and γ-secretase in lipid rafts from the three brain areas, in control and AD brains. We have observed that while the stage of the disease does not alter the level of association between APP-BACE and APP-PSEN1 in cerebellum, in the entorhinal and frontal cortices, the association between APP and BACE was considerably augmented when compared to the same areas in control lipid rafts. Conversely, physical association of APP and PSEN remained nearly constant between brain areas irrespective of disease stage. These findings are particularly relevant since β-cleavage of APP by BACE1 is the rate-limiting obligatory event, in the amyloidogenic pathway [6, 179, 180]. From a holistic perspective, the convergence of APP and BACE to lipid rafts, allows a closer interaction between the two proteins facilitating β-cleavage of AβPP and eventually Aβ production [6, 179–182]. These observations point to the existence of homeostatic mechanisms precluding their unabated convergence under non-pathological conditions. In agreement, in a recent study in cultured hippocampal neurons, specific trafficking strategies that limit APP/BACE-1 proximity in has been demonstrated under physiologic states [162], therefore limiting amyloidogenesis. However, in this later study, disturbing raft architecture by moderate (but not severe) reduction of cholesterol levels increase Aβ production by enhancing BACE1 and APP interaction [161, 162, 164]. Our results in human brain lipid rafts, agrees with this finding that moderate reduction in cholesterol facilitates convergence pathways that routes APP and BACE to lipid rafts. However, the most important factors in triggering this convergence are the reduction in LCPUFA and the increased proportions of saturates/n3 and phospholipids/cholesterol in lipid rafts from entorhinal and frontal cortices, which, as we have showed before, gives rise to more liquid-ordered microdomains, likely stabilizing the interaction of AβPP and BACE1. In this sense, lipids can build a physical boundary between domains, circumscribing the β-secretase-APP complex within the lipid raft domain, where the pool of γ-secretase resides, thus favoring the sequential amyloidogenic cleavage of APP [183].

On the other hand, plasmalogens, membrane glycerophospholipids abundant neuronal lipids, have also been associated to AD. Reduced levels of these brain-specific lipids have been reported in AD brains [12, 184]. This is relevant for three main reasons: first plasmalogens (particularly plasmenyl-ethanolamine, PlsEtn) act as neuronal depots for essential LCPUFA in the brain and structural determinants of acyl chains packing and membrane order [23]; Second because the oxidative products of plasmalogens are unable to further propagate lipid peroxidation, and essential factor in triggering AD, thus plasmalogens may terminate lipid oxidation [185] and third, because they might have direct effect on the production of Aβ by inhibiting activity of γ-secretase [184].

Of particular interest is the fact that the normal aging brain undergoes a set of lipid alterations in lipid rafts collectively termed "lipid raft aging" [53, 94, 135, 151, 152, 168]. Changes affect levels of sphingomyelin, sulfatides and cerebrosides, LCPUFA, plasmalogens, phosphatidylinositol, gangliosides, and total neutral lipids (mainly cholesterol and sterol esters). Further, relevant relationships between main fatty acids and/or lipid classes detected in younger subjects, either disappeared or they occurred in the opposite direction [157]. Noticeably, these changes are mostly subtle but follow the same trend observed in early stages of AD. "Lipid raft aging" also involves changes in unsaturation and peroxidability indexes though they are significantly less severe than those reported in AD cortex [56, 57], and do not cause significant biophysical alterations of raft membranes. The significant reduction in peroxidability indexes observed in early stages of AD (reflecting the important reduction of LCPUFA in both raft and non-raft domains), and especially during lipid raft aging, is strongly indicative that oxidative stress and exhaustion of antioxidant systems are an essential part of AD neurodegeneration.

Interestingly, "lipid raft aging" exhibits clear gender differences and appear to be more pronounced in women, especially in older postmenopausal women [168], which strengthens a role for ovarian hormones in AD development. Indeed, according to the Alzheimer's Association [186] women have 2-fold greater lifetime risk of developing AD. Though still incompletely understood, it seems clear that menopausal transition and decline in estrogen adversely affect brain metabolism [187, 188].

Overall, the evidence accumulated point to a complex cocktail of factors, either endogenous and/or environmental, affecting lipid raft physiology and stability as paramount events in trespassing the thin borderline that separates normal and pathological aging [158].

6. Conclusions

In summary, we may conclude that lipid rafts are the neurobiological locus for the wealth of alterations involved in the molecular pathophysiology of Alzheimer's disease. Severe changes in the lipid matrix of lipid rafts represent the seminal event in the pathogenesis of Alzheimer's disease. These early changes, that selectively affect cortical structures altered in AD, have a profound impact on physicochemical properties of lipid raft which serves a favorable environment for the abnormal neuronal physiology, especially for the interaction of secretases and APP to trigger the amyloidogenic processing of APP and amyloid burden. This review argues in favor of lipid rafts dyshomeostasis representing a foundational effect on the onset and progression of this devasting disease, and opens the possibility for new pharmacological approaches and therapeutic windows to halt the initiation of this neurodegenerative disease.

Acknowledgements

Supported by grants SAF2017-84454-R from Ministerio de Economía y Competitividad (MINECO, Spain) and ProID2020010075 from ACIISI (Canary Islands, Spain). I wish to express my deepest gratitude to Mrs. Doctoranda Concha for invaluable help during the elaboration of the final manuscript. Without her altruistic help, this article would not be ready on time. I wish to dedicate this article to the loving memory of my mother, Quina, who fade away in the realm of this unsparing disease. Funding from University of La Laguna and the Canary Council of Economy, Knowledge and Employment, CEI Program.

Author details

Mario Díaz[1,2,3*] and Raquel Marin[2,4]

1 Department of Biología Animal, Universidad de La Laguna, La Laguna, Tenerife, Spain

2 Unidad Asociada ULL-CSIC "Fisiología y Biofísica de la Membrana Celular en Enfermedades Neurodegenetarivas y Tumorales", La Laguna, Tenerife, Spain

3 Instituto Universitario de Neurociencia Cognitiva, Tenerife, Spain

4 School of Medicine, Universidad de La Laguna, La Laguna, Tenerife, Spain

*Address all correspondence to: madiaz@ull.es

References

[1] Pike CJ. Sex and the development of Alzheimer's disease. Journal of Neuroscience Research. 2017;**95**:671-680 https://doi.org/10.1002/jnr.23827

[2] Tsui-Pierchala BA, Encinas M, Milbrandt J, Johnson EM. Lipid rafts in neuronal signaling and function. Trends in Neurosciences. 2002;**25**:412-417 https://doi.org/10.1016/S0166-2236(02)02215-4

[3] Allen JA, Halverson-Tamboli RA, Rasenick MM. Lipid raft microdomains and neurotransmitter signalling. Nature Reviews. Neuroscience. 2007;**8**:128-140 https://doi.org/10.1038/nrn2059

[4] Michel V, Bakovic M. Lipid rafts in health and disease. Biology of the Cell. 2007;**99**:129-140 https://doi.org/10.1042/bc20060051

[5] Fantini J, Garmy N, Mahfoud R, Yahi N. Lipid rafts: Structure, function and role in HIV, Alzheimer's and prion diseases. Expert Reviews in Molecular Medicine. 2002;**4**:1-22 https://doi.org/10.1017/S1462399402005392

[6] Hicks DA, Nalivaeva NN, Turner AJ. Lipid rafts and Alzheimer's disease: Protein-lipid interactions and perturbation of signaling. Frontiers in Physiology. 2012:1-40 https://doi.org/10.3389/fphys.2012.00189

[7] Ewers H, Helenius A. Lipid-mediated endocytosis. Cold Spring Harbor Perspectives in Biology. 2011;**10**:1-14

[8] Suzuki T, Suzuki Y. Virus infection and lipid rafts. Biological & Pharmaceutical Bulletin. 2006;**29**:1538-1541 https://doi.org/10.1248/bpb.29.1538

[9] Marsh M, Helenius A. Virus entry: Open sesame. Cell. 2006;**124**:729-740 https://doi.org/10.1016/j.cell.2006.02.007

[10] Di Paolo G, Kim TW. Linking lipids to Alzheimer's disease: Cholesterol and beyond. Nature Reviews. Neuroscience. 2011;**12**:284-296 https://doi.org/10.1038/nrn3012

[11] Posse de Chaves E, Sipione S. Sphingolipids and gangliosides of the nervous system in membrane function and dysfunction. FEBS Letters. 2010;**584**:1748-1759 https://doi.org/10.1016/j.febslet.2009.12.010

[12] Sonnino S, Prinetti A. Membrane domains and the "lipid raft" concept. Current Medicinal Chemistry. 2012;**20**:4-21 https://doi.org/10.2174/09298673130103

[13] Sonnino S, Mauri L, Chigorno V, Prinetti A. Gangliosides as components of lipid membrane domains. Glycobiology. 2006;**17**:1-13

[14] Cantú L, Del Favero E, Sonnino S, Prinetti A. Gangliosides and the multiscale modulation of membrane structure. Chemistry and Physics of Lipids. 2011;**164**:796-810 https://doi.org/10.1016/j.chemphyslip.2011.09.005

[15] Kolter T. Ganglioside Biochemistry. ISRN Biochem. 2012;**2012**:1-36 https://doi.org/10.5402/2012/506160

[16] Chinnapen DJF, Hsieh WT, te Welscher YM, Saslowsky DE, Kaoutzani L, Brandsma E, et al. Lipid sorting by ceramide structure from plasma membrane to ER for the cholera toxin receptor ganglioside GM1. Developmental Cell. 2012;**23**:573-586 https://doi.org/10.1016/j.devcel.2012.08.002

[17] Matsuzaki K, Kato K, Yanagisawa K. Aβ polymerization through interaction with membrane gangliosides. Biochim Biophys Acta - Mol Cell Biol Lipids. 2010;**1801**:868-877 https://doi.org/10.1016/j.bbalip.2010.01.008

[18] Farooqui AA. Lipid mediators and their metabolism in the nucleus: Implications for Alzheimer's Disease. Journal of Alzheimer's Disease. 2012;**30**:163-178 https://doi.org/10.3233/JAD-2011-111085

[19] Hannun YA, Obeid LM. Principles of bioactive lipid signalling: Lessons from sphingolipids. Nature Reviews. Molecular Cell Biology. 2008;**9**:139-150 https://doi.org/10.1038/nrm2329

[20] Di Paolo G, De Camilli P. Phosphoinositides in cell regulation and membrane dynamics. Nature. 2006;**443**:651-657 https://doi.org/10.1038/nature05185

[21] Asatryan A, Bazan NG. Molecular mechanisms of signaling via the docosanoid neuroprotectin D1 for cellular homeostasis and neuroprotection. The Journal of Biological Chemistry. 2017;**292**:12390-12397 https://doi.org/10.1074/jbc.R117.783076

[22] Palacios-Pelaez R, Lukiw WJ, Bazan NG. Omega-3 essential fatty acids modulate initiation and progression of neurodegenerative disease. Molecular Neurobiology. 2010;**41**:367-374 https://doi.org/10.1007/s12035-010-8139-z

[23] Nagan N, Zoeller RA. Plasmalogens: Biosynthesis and functions. vol. 40. 2001. https://doi.org/10.1016/S0163-7827(01)00003-0.

[24] Ramstedt B, Slotte JP. Membrane properties of sphingomyelins. FEBS Letters. 2002;**531**:33-37 https://doi.org/10.1016/S0014-5793(02)03406-3

[25] Ramstedt B, Slotte JP. Sphingolipids and the formation of sterol-enriched ordered membrane domains. Biochim Biophys Acta - Biomembr. 2006;**1758**:1945-1956 https://doi.org/10.1016/j.bbamem.2006.05.020

[26] Xu X, London E. The effect of sterol structure on membrane lipid domains reveals how cholesterol can induce lipid domain formation. Biochemistry. 2000;**39**:843-849 https://doi.org/10.1021/bi992543v

[27] Barenholz Y, Thompson TE. Sphingomyelins in bilayers and biological membranes. Biochim Biophys Acta - Biomembr. 1980;**604**:129-158 https://doi.org/10.1016/0005-2736(80)90572-6

[28] Simons K, Sampaio JL. Membrane organization and lipid rafts. Cold Spring Harbor Perspectives in Biology. 2011;**3**:1-17 https://doi.org/10.1101/cshperspect.a004697

[29] Heberle FA, Feigenson GW. Phase separation in lipid membranes. Cold Spring Harbor Perspectives in Biology. 2011;**3**:1-13 https://doi.org/10.1101/cshperspect.a004630

[30] Simons K, Ikonen E. Functional rafts in cell membranes. Nature. 1997;**387**:569-572 https://doi.org/10.1038/42408

[31] Brown DA, London E. Structure and function of sphingolipid- and cholesterol-rich membrane rafts. The Journal of Biological Chemistry. 2000;**275**:17221-17224 https://doi.org/10.1074/jbc.R000005200

[32] Lingwood D, Simons K. Lipid Rafts As a Membrane- Organizing Principle 2010:46-51.

[33] Foster LJ, Chan QWT. Lipid raft proteomics: More than just detergent-resistant membranes. Sub-Cellular Biochemistry. 2007;**43**:35-47 https://doi.org/10.1007/978-1-4020-5943-8_4

[34] Shah A, Chen D, Boda AR, Foster LJ, Davis MJ, RaftProt HMM. Mammalian lipid raft proteome database. Nucleic Acids Research. 2015;**43**:D335-D338 https://doi.org/10.1093/nar/gku1131

[35] Sillence DJ. New insights into glycosphingolipid functions-storage,

lipid rafts, and translocators. International Review of Cytology. 2007;**262**:151-189 https://doi.org/10.1016/S0074-7696(07)62003-8

[36] Imran M, Mahmood S. An overview of human prion diseases. Virology Journal. 2011;**8**:1-9 https://doi.org/10.1186/1743-422X-8-559

[37] Wulf MA, Senatore A, Aguzzi A. The biological function of the cellular prion protein: An update. BMC Biology. 2017;**15**:1-13 https://doi.org/10.1186/s12915-017-0375-5

[38] Taylor DR, Hooper NM. The prion protein and lipid rafts (Review). Molecular Membrane Biology. 2006;**23**:89-99 https://doi.org/10.1080/09687860500449994

[39] Jiang H, Zhang X, Chen X, Aramsangtienchai P, Tong Z, Lin H. Protein Lipidation: Occurrence, mechanisms, biological functions, and enabling technologies. Chemical Reviews. 2018;**118**:919-988 https://doi.org/10.1021/acs.chemrev.6b00750

[40] Sato I, Obata Y, Kasahara K, Nakayama Y, Fukumoto Y, Yokoyama KK, et al. Differential trafficking of Src, Lyn, yes and Fyn is specified by the state of palmitoylation in the SH4 domain. Journal of Cell Science. 2009;**122**:965-975 https://doi.org/10.1242/jcs.034843

[41] Lee H, Woodman SE, Engelman JA, Volonte D, Galbiati F, Kaufman HL, et al. Palmitoylation of Caveolin-1 at a single site (Cys-156) controls its coupling to the c-Src tyrosine kinase: Targeting of dually acylated molecules (Gpi-linked, transmembrane, or cytoplasmic) to caveolae effectively uncouples c-Src and caveolin-1 (Tyr-14). The Journal of Biological Chemistry. 2001;**276**:35150-35158 https://doi.org/10.1074/jbc.M104530200

[42] Xu N, Shen N, Wang XX, Jiang S, Xue B, Li CJ. Protein prenylation and human diseases: A balance of protein farnesylation and geranylgeranylation. Science China. Life Sciences. 2015;**58**:328-335 https://doi.org/10.1007/s11427-015-4836-1

[43] Langhorst MF, Reuter A, Stuermer CAO. Scaffolding microdomains and beyond: The function of reggie/flotillin proteins. Cellular and Molecular Life Sciences. 2005;**62**:2228-2240 https://doi.org/10.1007/s00018-005-5166-4

[44] Okamoto T, Schlegel A, Scherer PE, Lisanti MP. Caveolins, a family of scaffolding proteins for organizing «preassembled signaling complexes» at the plasma membrane. The Journal of Biological Chemistry. 1998;**273**:5419-5422 https://doi.org/10.1074/jbc.273.10.5419

[45] Collins BM, Davis MJ, Hancock JF, Parton RG. Structure-based reassessment of the caveolin signaling model: Do caveolae regulate signaling through caveolin-protein interactions? Developmental Cell. 2012;**23**:11-20 https://doi.org/10.1016/j.devcel.2012.06.012

[46] Levental I, Lingwood D, Grzybek M, Coskun Ü, Simons K. Palmitoylation regulates raft affinity for the majority of integral raft proteins. Proceedings of the National Academy of Sciences of the United States of America. 2010;**107**:22050-22054 https://doi.org/10.1073/pnas.1016184107

[47] Couet J, Sargiacomo M, Lisanti MP. Interaction of a receptor tyrosine kinase, EGF-R, with caveolins. Caveolin binding negatively regulates tyrosine and serine/threonine kinase activities. The Journal of Biological Chemistry. 1997;**272**:30429-30438 https://doi.org/10.1074/jbc.272.48.30429

[48] Morrow IC, Rea S, Martin S, Prior IA, Prohaska R, Hancock JF, et al.

Flotillin-1/reggie-2 traffics to surface raft domains via a novel Golgi-independent pathway. Identification of a novel membrane targeting domain and a role for palmitoylation. The Journal of Biological Chemistry. 2002;**277**:48834-48841 https://doi.org/10.1074/jbc.M209082200

[49] Epand RM. Cholesterol and the interaction of proteins with membrane domains. Progress in Lipid Research. 2006;**45**:279-294 https://doi.org/10.1016/j.plipres.2006.02.001

[50] Yeagle PL. Non-covalent binding of membrane lipids to membrane proteins. Biochim Biophys Acta - Biomembr. 2014;**1838**:1548-1559 https://doi.org/10.1016/j.bbamem.2013.11.009

[51] Suzuki T. Lipid rafts at postsynaptic sites: Distribution, function and linkage to postsynaptic density. Neuroscience Research. 2002;**44**:1-9 https://doi.org/10.1016/S0168-0102(02)00080-9

[52] Suzuki T, Zhang J, Miyazawa S, Liu QF, A MR, Yao WD. Association of membrane rafts and postsynaptic density: Proteomics, biochemical, and ultrastructural analyses. Journal of Neurochemistry. 2011;**119**:64-77 https://doi.org/10.1111/j.1471-4159.2011.07404.x

[53] Suzuki T, Du F, Tian QB, Zhang J, Endo S. Ca2+/calmodulin-dependent protein kinase IIα clusters are associated with stable lipid rafts and their formation traps PSD-95. Journal of Neurochemistry. 2008;**104**:596-610 https://doi.org/10.1111/j.1471-4159.2007.05035.x

[54] García-Sáez AJ, Schwille P. Stability of lipid domains. FEBS Letters. 2010;**584**:1653-1658 https://doi.org/10.1016/j.febslet.2009.12.036

[55] Marin R, Fabelo N, Martín V, Garcia-Esparcia P, Ferrer I, Quinto-Alemany D, et al. Anomalies occurring in lipid profiles and protein distribution in frontal cortex lipid rafts in dementia with Lewy bodies disclose neurochemical traits partially shared by Alzheimer's and Parkinson's diseases. Neurobiology of Aging. 2017;**49**:52-59 https://doi.org/10.1016/j.neurobiolaging.2016.08.027

[56] Fabelo N, Martín V, Marín R, Santpere G, Aso E, Ferrer I, et al. Evidence for premature lipid raft aging in APP/PS1 double-transgenic mice, a model of familial Alzheimer disease. Journal of Neuropathology and Experimental Neurology. 2012;**71**:868-881 https://doi.org/10.1097/NEN.0b013e31826be03c

[57] Díaz M, Luis-Amaro AC, Barreto DR, Casañas-Sánchez V, Pérez JA, Marin R. Lipostatic mechanisms preserving cerebellar lipids in MPTP-treated mice: Focus on membrane microdomains and lipid-related gene expression. Frontiers in Molecular Neuroscience. 2019;**12** https://doi.org/10.3389/fnmol.2019.00093

[58] Fabelo N, Martin V, Santpere G, Marín R, Torrent L, Ferrer I, et al. Severe alterations in lipid composition of frontal cortex lipid rafts from Parkinson's Disease and incidental Parkinson's Disease. Molecular Medicine. 2011;**17**(1) https://doi.org/10.2119/molmed.2011.00119

[59] Diaz ML, Fabelo N, Marín R. Genotype-induced changes in biophysical properties of frontal cortex lipid raft from APP/PS1 transgenic mice. Frontiers in Physiology. 2012;**3**:1-11 https://doi.org/10.3389/fphys.2012.00454

[60] Díaz M, Fabelo N, Martín V, Ferrer I, Gómez T, Marín R. Biophysical alterations in lipid rafts from human cerebral cortex associate with increased BACE1/AβPP interaction in early stages of Alzheimer's disease. J Alzheimer's Dis. 2014;**43**:1185-1198 https://doi.org/10.3233/JAD-141146

[61] Harroun TA, Katsaras J, Wassall SR. Cholesterol is found to reside in the center of a polyunsaturated lipid membrane. Biochemistry. 2008;**47**:7090-7096 https://doi.org/10.1021/bi800123b

[62] Schlessinger J. Cell signaling by receptor tyrosine kinases. Encycl Earth Sci Ser. 2000;**103**:211-225 https://doi.org/10.1007/978-90-481-2642-2_16

[63] Canu N, Amadoro G, Triaca V, Latina V, Sposato V, Corsetti V, et al. The intersection of NGF/TrkA signaling and amyloid precursor protein processing in Alzheimer's disease neuropathology. International Journal of Molecular Sciences. 2017;**18**:1-17 https://doi.org/10.3390/ijms18061319

[64] Marin R, Díaz M, Alonso R, Sanz A, Arévalo MA, Garcia-Segura LM. Role of estrogen receptor α in membrane-initiated signaling in neural cells: Interaction with IGF-1 receptor. The Journal of Steroid Biochemistry and Molecular Biology. 2009;**114**:2-7 https://doi.org/10.1016/j.jsbmb.2008.12.014

[65] Freeman MR, Cinar B, Kim J, Mukhopadhyay NK, Di Vizio D, Adam RM, et al. Transit of hormonal and EGF receptor-dependent signals through cholesterol-rich membranes. Steroids. 2007;**72**:210-217 https://doi.org/10.1016/j.steroids.2006.11.012

[66] Sil S, Periyasamy P, Thangaraj A, Chivero ET, Buch S. PDGF/PDGFR axis in the neural systems. Molecular Aspects of Medicine. 2018;**62**:63-74 https://doi.org/10.1016/j.mam.2018.01.006

[67] Liu P, Ying Y, Ko YG, Anderson RGW. Localization of platelet-derived growth factor-stimulated phosphorylation cascade to caveolae. The Journal of Biological Chemistry. 1996;**271**:10299-10303 https://doi.org/10.1074/jbc.271.17.10299

[68] Encinas M, Tansey MG, Tsui-PierchalaBA,ComellaJX,MilbrandtJ, Johnson EM. c-Src is required for glial cell line-derived neurotrophic factor (GDNF) family ligand-mediated neuronal survival via a phosphatidylinositol-3 kinase (PI-3K)-dependent pathway. The Journal of Neuroscience. 2001;**21**:1464-1472 https://doi.org/10.1523/jneurosci.21-05-01464.2001

[69] Parsons SJ, Parsons JT. Src family kinases, key regulators of signal transduction. Oncogene. 2004;**23**:7906-7909 https://doi.org/10.1038/sj.onc.1208160

[70] Superti-Furga G. Regulation of the Src protein tyrosine kinase. FEBS Letters. 1995;**369**:62-66 https://doi.org/10.1016/0014-5793(95)00636-N

[71] Kaplan DR, Miller FD. Neurotrophin signal transduction in the nervous system. Current Opinion in Neurobiology. 2000;**10**:381-391 https://doi.org/10.1016/S0959-4388(00)00092-1

[72] Kalia L V, Kalia SK, Salter MW. doi:10.1016/S1474-4422(08)70165-0. Lancet Neurol 2008;7:1-14.

[73] Beal M. The role of excitotoxicity in neurological disease. Current Opinion in Neurobiology. 1992;2:657-662

[74] Doshi S, Lynch DR. Calpain and the glutamatergic synapse. Front Biosci - Sch 2009;1 S:466-76. https://doi.org/10.2741/e38.

[75] Hossain MI, Roulston CL, Kamaruddin MA, Chu PWY, Ng DCH, Dusting GJ, et al. A truncated fragment of Src protein kinase generated by calpain-mediated cleavage is a mediator of neuronal death in excitotoxicity. The Journal of Biological Chemistry. 2013;**288**:9696-9709 https://doi.org/10.1074/jbc.M112.419713

[76] Mahaman YAR, Huang F, Kessete Afewerky H, Salissou MTM, Ghose B,

Wang X. Involvement of calpain in the neuropathogenesis of Alzheimer's disease. Medicinal Research Reviews. 2019;**39**:608-630 https://doi.org/10.1002/med.21534

[77] Garcia-Segura LM, Azcoitia I, DonCarlos LL. Neuroprotection by estradiol. vol. 63. 2001. https://doi.org/10.1016/S0301-0082(00)00025-3.

[78] Marin R, Guerra B, Alonso R, Ramirez C, Diaz M. Estrogen activates classical and alternative mechanisms to orchestrate neuroprotection. Current Neurovascular Research. 2005;**2**:287-301 https://doi.org/10.2174/156720205774322629

[79] Pedram A, Razandi M, Sainson RCA, Kim JK, Hughes CC, Levin ER. A conserved mechanism for steroid receptor translocation to the plasma membrane. The Journal of Biological Chemistry. 2007;**282**:22278-22288 https://doi.org/10.1074/jbc.M611877200

[80] Marin R. Signalosomes in the brain: Relevance in the development of certain neuropathologies such as Alzheimer ' s disease. Frontiers in Physiology. 2011;**2**:1-4 https://doi.org/10.3389/fphys.2011.00023

[81] Boonyaratanakornkit V, Edwards DP. Receptor mechanisms mediating non-genomic actions of sex steroids. Seminars in Reproductive Medicine. 2007;**25**:139-153 https://doi.org/10.1055/s-2007-973427

[82] Maudsley S, Chadwick W, Brenneman R, Martin B. Complex and multidimensional lipid raft alterations in a murine model of Alzheimer's disease. International Journal of Alzheimer's Disease. 2010;**2010**:1-56 https://doi.org/10.4061/2010/604792

[83] Chen Y, Fu AKY, Ip NY. Synaptic dysfunction in Alzheimer's disease: Mechanisms and therapeutic strategies. Pharmacology & Therapeutics. 2019;**195**:186-198 https://doi.org/10.1016/j.pharmthera.2018.11.006

[84] Holtzman DM, Morris JC, Goate AM. Science Translational Medicine Volume 3 issue 77 2011. doi 10.1126%2Fscitranslmed.3002369] Holtzman, D. M.; Morris, J. C.; Goate, A. M. -- Alzheimer's Disease- The Challenge of the Second Century.pdf 2011;3.

[85] Marin R, Ramirez CM, González M, Gonzalez Muñoz E, Zorzano A, Camps M, et al. Voltage-dependent anion channel (VDAC) participates in amyloid beta-induced toxicity and interacts with plasma membrane estrogen receptor a in septal and hippocampal neurons. Molecular Membrane Biology. 2007;**24**:148-160 https://doi.org/10.1080/09687860601055559

[86] Ramírez CM, González M, Díaz M, Alonso R, Ferrer I, Santpere G, et al. Molecular and cellular neuroscience VDAC and ER α interaction in caveolae from human cortex is altered in Alzheimer ' s disease. Molecular and Cellular Neurosciences. 2009;**42**:172-183 https://doi.org/10.1016/j.mcn.2009.07.001

[87] Akanda N, Tofighi R, Brask J, Tamm C, Elinder F, Ceccatelli S. Voltage-dependent anion channels (VDAC) in the plasma membrane play a critical role in apoptosis in differentiated hippocampal neurons but not in neural stem cells. Cell Cycle. 2008;**7**:3225-3234 https://doi.org/10.4161/cc.7.20.6831

[88] Le Roy C, Wrana JL. Clathrin- and non-clathrin-mediated endocytic regulation of cell signalling. Nature Reviews. Molecular Cell Biology. 2005;**6**:112-126 https://doi.org/10.1038/nrm1571

[89] Lajoie P, Nabi IR. Regulation of raft-dependent endocytosis. Journal of Cellular and Molecular Medicine. 2007;**11**:644-653 https://doi.org/10.1111/j.1582-4934.2007.00083.x

[90] Martens JR, O'Connell K, Tamkun M. Targeting of ion channels to membrane microdomains: Localization of K V channels to lipid rafts. Trends in Pharmacological Sciences. 2004;**25**:16-21 https://doi.org/10.1016/j.tips.2003.11.007

[91] Hibino H, Kurachi Y. Distinct detergent-resistant membrane microdomains (lipid rafts) respectively harvest K+ and water transport systems in brain astroglia. The European Journal of Neuroscience. 2007;**26**:2539-2555 https://doi.org/10.1111/j.1460-9568.2007.05876.x

[92] Murphy MP, Iii HL. NIH Public Access. J Alzheimer's Dis 2010;19:1-17. https://doi.org/10.3233/JAD-2010-1221. Alzheimer.

[93] Levitan I, Fang Y. Avia Rosenhouse-Dantsker VR. Cholesterol and Ion Channels. vol. 2010:51 https://doi.org/10.1007/978-90-481-8622-8

[94] O'Connell KMS, Tamkun MM. Targeting of voltage-gated potassium channel isoforms to distinct cell surface microdomains. Journal of Cell Science. 2005;**118**:2155-2166 https://doi.org/10.1242/jcs.02348

[95] Limpert AS, Karlo JC, Landreth GE. Nerve growth factor stimulates the concentration of TrkA within lipid rafts and extracellular signal-regulated kinase activation through c-Cbl-associated protein. Molecular and Cellular Biology. 2007;**27**:5686-5698 https://doi.org/10.1128/mcb.01109-06

[96] Head BP, Patel HH, Insel PA. Interaction of membrane/lipid rafts with the cytoskeleton: Impact on signaling and function. Membrane/lipid rafts, mediators of cytoskeletal arrangement and cell signaling. Biochim Biophys Acta - Biomembr. 2014;**1838**:532-545 https://doi.org/10.1016/j.bbamem.2013.07.018

[97] Ritchie K, Iino R, Fujiwara T, Murase K, Kusumi A. The fence and picket structure of the plasma membrane of live cells as revealed by single molecule techniques (Review). Molecular Membrane Biology. 2003;**20**:13-18 https://doi.org/10.1080/0968768021000055698

[98] Goswami D, Gowrishankar K, Bilgrami S, Ghosh S, Raghupathy R, Chadda R, et al. Nanoclusters of GPI-anchored proteins are formed by cortical actin-driven activity. Cell. 2008;**135**:1085-1097 https://doi.org/10.1016/j.cell.2008.11.032

[99] Arnold SE, Hyman BT, Flory J, Damasio AR, Van Hoesen GW. The topographical and neuroanatomical distribution of neurofibrillary tangles and neuritic plaques in the cerebral cortex of patients with alzheimer's disease. Cerebral Cortex. 1991;**1**:103-116 https://doi.org/10.1093/cercor/1.1.103

[100] Fjell AM, McEvoy L, Holland D, Dale AM, Walhovd KB. What is normal in normal aging? Effects of aging, amyloid and Alzheimer's disease on the cerebral cortex and the hippocampus. Progress in Neurobiology. 2014;**117**:20-40 https://doi.org/10.1016/j.pneurobio.2014.02.004

[101] Hardy J, Disease HGA' s. The amyloid Cascade hypothesis. Science (80-). 1992;**256**:184-185 https://doi.org/10.1126/science.1566067

[102] Vassar R, Bennett BD, Babu-Khan S, Kahn S, Mendiaz EA, Denis P, et al. β-Secretase cleavage of Alzheimer's amyloid precursor protein by the transmembrane aspartic protease

BACE. Science (80-) 1999;286:735-41. https://doi.org/10.1126/science.286.5440.735.

[103] Edbauer D, Winkler E, Regula JT, Pesold B, Steiner H, Haass C. Reconstitution of γ-secretase activity. Nature Cell Biology. 2003;**5**:486-488 https://doi.org/10.1038/ncb960

[104] Finder VH, Glockshuber R. Amyloid-β aggregation. Neurodegenerative Diseases. 2007;**4**: 13-27 https://doi.org/10.1159/000100355

[105] Allinson TMJ, Parkin ET, Turner AJ, Hooper NM. ADAMs family members As amyloid precursor protein α-secretases. Journal of Neuroscience Research. 2003;**352**:342-352

[106] Yuan X, Sun S, Tan C, Yu J, Tan L. The role of ADAM10 in Alzheimer's Disease. Journal of Alzheimer's Disease. 2017;**58**:303-322 https://doi.org/10.3233/JAD-170061

[107] Berthelot K, Cullin C, Lecomte S. What does make an amyloid toxic: Morphology, structure or interaction with membrane? Biochimie. 2013;**95**:12-19 https://doi.org/10.1016/j.biochi.2012.07.011

[108] Lee VYM, Goedert M, Trojanowski JQ. Neurodegenerative tauopathies. Annual Review of Neuroscience. 2001;**24**:1121-1161

[109] Braak H, Braak E. Neuropathological stageing of Alzheimer-related changes. Acta Neuropathologica. 1991;**82**:239-259 https://doi.org/10.1109/ICINIS.2015.10

[110] Bertram L, Tanzi RE. The genetic epidemiology of neurodegenerative disease. The Journal of Clinical Investigation. 2005;**115**:1449-1457 https://doi.org/10.1172/JCI24761

[111] Small SA, Duff K. Linking Aβ and tau in late-onset Alzheimer's Disease: A dual pathway hypothesis. Neuron. 2008;**60**:534-542 https://doi.org/10.1016/j.neuron.2008.11.007

[112] Chen TY, Liu PH, Ruan CT, Chiu L, Kung FL. The intracellular domain of amyloid precursor protein interacts with flotillin-1, a lipid raft protein. Biochemical and Biophysical Research Communications. 2006;**342**:266-272 https://doi.org/10.1016/j.bbrc.2006.01.156

[113] Wojsiat J, Zoltowska KM, Laskowska-Kaszub K, Wojda U, Zhang C, Wang K, et al. Targeting of voltage-gated potassium channel isoforms to distinct cell surface microdomains. The Journal of Biological Chemistry. 2010;**3**:1-22 https://doi.org/10.1242/jcs.02348

[114] Hattori C, Asai M, Onishi H, Sasagawa N, Hashimoto Y, Saido TC, et al. BACE1 interacts with lipid raft proteins. Journal of Neuroscience Research. 2006;**84**:912-917 https://doi.org/10.1002/jnr

[115] Cordy JM, Hussain I, Dingwall C, Hooper NM, Turner AJ. Exclusively targeting β-secretase to lipid rafts by GPI-anchor addition up-regulates β-site processing of the amyloid precursor protein. Proceedings of the National Academy of Sciences of the United States of America. 2003;**100**:11735-11740 https://doi.org/10.1073/pnas.1635130100

[116] Hur JY, Welander H, Behbahani H, Aoki M, Frånberg J, Winblad B, et al. Active γ-secretase is localized to detergent-resistant membranes in human brain. The FEBS Journal. 2008;**275**:1174-1187 https://doi.org/10.1111/j.1742-4658.2008.06278.x

[117] Cheng H, Vetrivel KS, Drisdel RC, Meckler X, Gong P, Leem JY, et al. S-Palmitoylation of γ-secretase subunits nicastrin and aph-1. The Journal of

Biological Chemistry. 2009;**284**:1373-1384 https://doi.org/10.1074/jbc.M806380200

[118] Kapoor A, Hsu WM, Wang BJ, Wu GH, Lin TY, Lee SJ, et al. Caveolin-1 regulates γ-secretase-mediated AβPP processing by modulating spatial distribution of γ-secretase in membrane. J Alzheimer's Dis. 2010;**22**:423-442 https://doi.org/10.3233/JAD-2010-100531

[119] Saijo E, Scheff SW, Telling GC. Unaltered prion protein expression in Alzheimer disease patients. Prion. 2011:109-116 https://doi.org/10.4161/pri.5.2.16355

[120] You H, Tsutsui S, Hameed S, Kannanayakal TJ, Chen L, Xia P. Aβ neurotoxicity depends on interactions between copper ions, prion protein, and N-methyl-D aspartate receptors. Proceedings of the National Academy of Sciences. 2012,**109**:1737-1742 https://doi.org/10.1073/pnas.1110789109

[121] Dietschy JM, Turley SD. Cholesterol metabolism in the. Brain 2001:105-112

[122] Marquer C, Devauges V, Cossec J, Duyckaerts C, Le S. Local cholesterol increase triggers amyloid precursor protein-Bace1 clustering in lipid rafts and rapid endocytosis. s. f. :1295-1305 https://doi.org/10.1096/fj.10-168633

[123] Kalvodova L, Kahya N, Schwille P, Ehehalt R, Verkade P, Drechsel D, et al. Lipids as modulators of proteolytic activity of BACE: Involvement of cholesterol, glycosphingolipids, and anionic phospholipids in vitro. The Journal of Biological Chemistry. 2005;**280**:36815-36823 https://doi.org/10.1074/jbc.M504484200

[124] Wahrle S, Das P, Nyborg AC, McLendon C, Shoji M, Kawarabayashi T, et al. Cholesterol-dependent γ-secretase activity in buoyant cholesterol-rich membrane microdomains. Neurobiology of Disease. 2002;**9**:11-23 https://doi.org/10.1006/nbdi.2001.0470

[125] Chang T-Y, Yamauchi Y, Hasan MT, Chang C. Cellular cholesterol homeostasis and Alzheimer ' s Disease. Journal of Lipid Research. 2017;**58**:603-650

[126] Xue-shan Z, Qi W, Zhong R, Li-hong P, Zhi- T, Zhi-sheng J, et al. Imbalanced cholesterol metabolism in Alzheimer's disease. Clinica Chimica Acta. 2016;**1**:107-114 https://doi.org/10.1016/j.cca.2016.02.024

[127] Shobab LA, Hsiung GR, Feldman HH. Cholesterol in Alzheimer's disease. Lancet. 2005;**4**:841-852

[128] Wood WG, Schroeder F, Igbavboa U, Avdulov NA, Chochina SV. Brain membrane cholesterol domains, aging and amyloid beta-peptides. Neurobiology of Aging. 2002;**23**:685-694

[129] Araki W, Tamaoka A. Amyloid beta-protein and lipid rafts: Focused on biogenesis and catabolism 3. A β BIOGENESIS AND LIPID RAFTS. 2015:314-324

[130] Kim J, Basak JM, Review HDM. The role of apolipoprotein E in Alzheimer's Disease. Neuron. 2009;**63**:287-303 https://doi.org/10.1016/j.neuron.2009.06.026

[131] Lambert DW. Molecular biology of the SARS-coronavirus. Mol Biol SARS-Coronavirus. 2010;**2**:1-328 https://doi.org/10.1007/978-3-642-03683-5

[132] Corder E, Saunders AM, Strittmatter WJ, Schmechel DE, Gaskell PC, Small GW, et al. Gene Dose of Apolipoprotein E Type 4 Allele and the Risk of Alzheimer's Disease in Late Onset Families. Science (80-) 1993;8:41-3.

[133] Riedel BC, Thompson PM, Brinton RD. Age, APOE and sex: Triad of risk of Alzheimer's disease. The Journal of Steroid Biochemistry and Molecular Biology. 2016;**160**:134-147 https://doi.org/10.1016/j.jsbmb.2016.03.012

[134] Bu G. Apolipoprotein E and its receptors in Alzheimer ' s disease: pathways, pathogenesis and therapy 2009;10:333-44. https://doi.org/10.1038/nrn2620.

[135] Kao YC, Ho PC, Tu YK, Jou IM, Tsai KJ. Lipids and alzheimer's disease. International Journal of Molecular Sciences. 2020;**21**:1-37 https://doi.org/10.3390/ijms21041505

[136] Mesa-Herrera F, Taoro-González L, Valdés-Baizabal C, Diaz M, Marín R. Lipid and lipid raft alteration in aging and neurodegenerative diseases: A window for the development of new biomarkers. International Journal of Molecular Sciences. 2019;**20** https://doi.org/10.3390/ijms20153810

[137] Marin R, Fabelo N, Fernández-Echevarría C, Canerina-AmaroA,Rodríguez-BarretoD, Quinto-Alemany D, et al. Lipid raft alterations in aged-associated Neuropathologies. Current Alzheimer Research. 2016;**13**:973-984 https://doi.org/10.2174/1567205013666160314150017

[138] Hartmann T, Kuchenbecker J. Grimm MOW. Alzheimer ' s disease: the lipid connection. 2007;**103**:159-170

[139] Huang TL. Omega-3 fatty acids, cognitive decline, and Alzheimer's disease: A critical review and evaluation of the literature. J Alzheimer's Dis. 2010;**21**:673-690 https://doi.org/10.3233/JAD-2010-090934

[140] Chew H, Solomon VA, Fonteh AN. Involvement of lipids in Alzheimer's Disease pathology and potential therapies. Frontiers in Physiology. 2020;**11**:1-28 https://doi.org/10.3389/fphys.2020.00598

[141] Prasad MR, Lovell MA, Yatin M, Dhillon H, Markesbery WR. Regional membrane phospholipid alterations in Alzheimer's Disease. Neurochemical Research. 1998;**23**:81-88

[142] Söderberg M, Edlund C, Kristensson K, Dallner G. Lipid compositions of different regions of the human brain during aging. Journal of Neurochemistry. 1990;**54**:415-423 https://doi.org/10.1111/j.1471-4159.1990.tb01889.x

[143] Soderberg M, Edlund C, Kristensson K, Dallner G. Fatty acid composition of brain phospholipids in aging and in Alzheimer ' s Disease. Lipids. 1991;**26**:421-425

[144] Soderberg M, Edlund C, Alafuzoff I, Kristensson K, Dallner G. Lipid composition in different regions of the brain in Alzheimer's Disease /senile dementia of Alzheimer's type. Journal of Neurochemistry. 1992:1646-1653

[145] Shaikh SR, Cherezov V, Caffrey M, Stillwell W, Wassall SR. Interaction of cholesterol with a docosahexaenoic acid-containing phosphatidylethanolamine: Trigger for microdomain/ raft formation? Biochemistry. 2003:12028-12037

[146] Uauy R, Hoffman DR, Peirano P, Birch DG, Birch EE. Essential fatty acids in visual and brain development. Lipids. 2001;**36**:885-895

[147] Calderon F, Kim H. Docosahexaenoic acid promotes neurite growth in hippocampal neurons 2004:979-88. https://doi.org/10.1111/j.1471-4159.2004.02520.x.

[148] Luo Y, Niu F, Sun Z, Cao W, Zhang X, Guan D, et al. Altered

expression of Aβ metabolism-associated molecules from d-galactose/AlCl3 induced mouse brain. Mechanisms of Ageing and Development. 2009;**130**:248-252 https://doi.org/10.1016/j.mad.2008.12.005

[149] Bazinet RP. Polyunsaturated fatty acids and their metabolites in brain function and disease 2014. https://doi.org/10.1038/nrn3820.

[150] Oster T, Pillot T. Docosahexaenoic acid and synaptic protection in Alzheimer's disease mice. BBA - Mol Cell Biol Lipids. 2010;**1801**:791-798 https://doi.org/10.1016/j.bbalip.2010.02.011

[151] Grimm MOW, Kuchenbecker J, Grosgen S, Burg VK, Hundsdorfer B, Rothhaar TL, et al. Docosahexaenoic acid reduces amyloid β production via multiple pleiotropic mechanisms. The Journal of Biological Chemistry. 2011;**286**:14028-14039 https://doi.org/10.1074/jbc.M110.182329

[152] Casañas-Sánchez V, Pérez JA, Fabelo N, Herrera-Herrera AV, Fernández C, Marín R, et al. Addition of docosahexaenoic acid, but not arachidonic acid, activates glutathione and thioredoxin antioxidant systems in murine hippocampal HT22 cells: Potential implications in neuroprotection. Journal of Neurochemistry. 2014;**131**:470-483 https://doi.org/10.1111/jnc.12833

[153] Casañas-sánchez V, Pérez JA, Fabelo N, Quinto-alemany D. Docosahexaenoic (DHA) modulates expression to ensure self-protection from oxidative damage in hippocampal. Cell. 2015;**6**:1-11 https://doi.org/10.3389/fphys.2015.00203

[154] He X, Huang Y, Li B, Gong CX, Schuchman EH. Deregulation of sphingolipid metabolism in Alzheimer's disease. Neurobiology of Aging. 2010;**31**:398-408 https://doi.org/10.1016/j.neurobiolaging.2008.05.010

[155] Díaz M, Marín R. Brain polyunsaturated lipids and neurodegenerative diseases. Nutraceuticals Funct. Foods Nat. Rem., 2013, p. 387-412.

[156] Droge W. Free radicals in the physiological control of cell function. pdf. Physiol Rev. 2002:47-95 https://doi.org/10.1152/physrev.00018.2001

[157] Valko M, Leibfritz D, Moncol J, Cronin MTD, Mazur M, Telser J. Free radicals and antioxidants in normal physiological functions and human disease. The International Journal of Biochemistry & Cell Biology. 2007;**39**:44-84 https://doi.org/10.1016/j.biocel.2006.07.001

[158] Catalá A. Díaz M. Impact of Lipid Peroxidation on the Physiology and Pathophysiology of Cell Membranes. 2017. DOI: https://doi.org/10.3389/978-2-88945-082-4

[159] Catalá A. Lipid peroxidation of membrane phospholipids generates hydroxy-alkenals and oxidized phospholipids active in physiological and/or pathological conditions. Chemistry and Physics of Lipids. 2009;**157**:1-11 https://doi.org/10.1016/j.chemphyslip.2008.09.004

[160] Fritz KS, Petersen DR. An overview of the chemistry and biology of reactive aldehydes. Free Radical Biology & Medicine. 2013;**59**:85-91 https://doi.org/10.1016/j.freeradbiomed.2012.06.025

[161] Martín V, Fabelo N, Santpere G, Puig B, Marín R, Ferrer I, et al. Lipid alterations in lipid rafts from Alzheimer's Disease human brain cortex. J Alzheimer's Dis. 2010;**19**:489-502 https://doi.org/10.3233/jad-2010-1242

[162] Stillwell W, Wassall SR. Docosahexaenoic acid: Membrane properties of a unique fatty acid. Chemistry and Physics of Lipids. 2003;**126**:1-27 https://doi.org/10.1016/S0009-3084(03)00101-4

[163] Kracun I, Rosner H, Drnovsek V, Vukelic Z, Cosovic C, Trbojevic-Cepe M, et al. Gangliosides in the human brain development and aging. Neurochemistry International. 1992;**20**:421-431 https://doi.org/10.1016/0197-0186(92)90057-X

[164] Ariga T, Mcdonald MP. Yu RK. Role of ganglioside metabolism in the pathogenesis of Alzheimer's disease - a review. Journal of Lipid Research. 2008;**49**:1157-1175 https://doi.org/10.1194/jlr.R800007-JLR200

[165] Molander-melin M, Blennow K, Bogdanovic N, Dellheden B, Fredman P. Structural membrane alterations in Alzheimer brains found to be associated with regional disease development ; increased density of gangliosides GM1 and GM2 and loss of cholesterol in detergent- resistant membrane domains. Journal of Neurochemistry. 2005:171-182 https://doi.org/10.1111/j.1471-4159.2004.02849.x

[166] Okada T, Ikeda K, Wakabayashi M, Ogawa M, Matsuzaki K. Formation of toxic A β (1 – 40) fibrils on GM1 ganglioside-containing membranes mimicking lipid rafts: Polymorphisms in Aβ (1 – 40) fibrils. Journal of Molecular Biology. 2008;**382**:1066-1074 https://doi.org/10.1016/j.jmb.2008.07.072

[167] Yamamoto N, Igbabvoa U, Shimada Y, Ohno-iwashita Y, Kobayashi M, Wood WG, et al. Accelerated A b aggregation in the presence of GM1-ganglioside-accumulated synaptosomes of aged apoE4-knock-in mouse brain. The FEBS Journal. 2004;**569**:135-139 https://doi.org/10.1016/j.febslet.2004.05.037

[168] Kakio A, Yano Y, Takai D, Kuroda Y, Matsumoto O, Kozutsumi Y, et al. Interaction between amyloid β-protein aggregates and membranes. Journal of Peptide Science. 2004;**10**:612-621 https://doi.org/10.1002/psc.570

[169] Lemkul JA, Bevan DR. Lipid composition influences the release of Alzheimer ' s amyloid b -peptide from membranes. Protein Science. 2011;**20**:1530-1545 https://doi.org/10.1002/pro.678

[170] Kracun I, Rosner H, V. D, Heffer-Lauc M, Cosovic C, Lauc G. Human brain gangliosides. Int J Dev Biol 1991;35:289-95.

[171] Fabelo N, Martín V, Marín R, Moreno D, Ferrer I, Díaz M. Altered lipid composition in cortical lipid rafts occurs at early stages of sporadic Alzheimer's disease and facilitates APP/BACE1 interactions. Neurobiology of Aging. 2014;**35**:1801-1812 https://doi.org/10.1016/j.neurobiolaging.2014.02.005

[172] Braak H, Braak E. Frequency of stages of Alzheimer-related lesions in different age categories. Neurobiology of Aging. 1997;**18**:377-379 https://doi.org/10.1016/S0197-4580(97)00051-1

[173] Han X, Holtzman DM, McKeel DW, Kelley J, Morris JC. Substantial sulfatide deficiency and ceramide elevation in very early Alzheimer's disease: Potential role in disease pathogenesis. Journal of Neurochemistry. 2002;**82**:809-818 https://doi.org/10.1046/j.1471-4159.2002.00997.x

[174] Matsuzaki K. Physicochemical interactions of amyloid β-peptide with lipid bilayers. Biochim Biophys Acta - Biomembr. 2007;**1768**:1935-1942 https://doi.org/10.1016/j.bbamem.2007.02.009

[175] Aso E, Lomoio S, López-gonzález I, Joda L, Carmona M, Fernández-yagüe N, et al. Amyloid Generation and

Dysfunctional Immunoproteasome Activation with Disease Progression in Animal Model of Familial Alzheimer ' s Disease 2012;22:636-53. https://doi.org/10.1111/j.1750-3639.2011.00560.x.

[176] Ferrer I. Early involvement of the cerebral cortex in Parkinson's disease: Convergence of multiple metabolic defects. Progress in Neurobiology. 2009;**88**:89-103 https://doi.org/10.1016/j.pneurobio.2009.02.004

[177] Serrano-pozo A, Frosch MP, Masliah E, Hyman BT, Holtzman DM, Mandelkow E, et al. Neuropathological alterations in Alzheimer Disease. Cold Spring Harbor Perspectives in Biology. 2011:1-23 https://doi.org/10.1101/cshperspect.a006189

[178] Braak H, Alafuzo I, Arzberger VT, Kretzschmar H, Del K. Staging of Alzheimer disease-associated neuro W brillary pathology using para Y n sections and immunocytochemistry 2006:389-404. https://doi.org/10.1007/s00401-006-0127-z.

[179] Askarova S, Yang X, Lee JC. Impacts of Membrane Biophysics in Alzheimer ' s Disease: From Amyloid Precursor Protein Processing to A β Peptide-Induced Membrane Changes 2011;2011. https://doi.org/10.4061/2011/134971.

[180] Dislich B, Lichtenthaler SF. The membrane-bound aspartyl protease BACE1: molecular and functional properties in Alzheimer ' s disease and beyond 2012;3:1-16. https://doi.org/10.3389/fphys.2012.00008.

[181] Abad-rodriguez J, Ledesma MD, Craessaerts K, Perga S, Medina M, Delacourte A, et al. Neuronal membrane cholesterol loss enhances amyloid peptide generation. The Journal of Cell Biology. 2004;**167**:953-960 https://doi.org/10.1083/jcb.200404149

[182] Das U, Scott DA, Ganguly A, Koo EH, Tang Y, Roy S. Activity-induced convergence of APP and BACE-1 in acidic microdomains via endocytosis-dependent pathway. Neuron. 2013;**79**:447-460 https://doi.org/10.1016/j.neuron.2013.05.035

[183] Kaether C, Haass C. A lipid boundary separates APP and secretases and limits amyloid β-peptide generation. The Journal of Cell Biology. 2004;**167**:809-812 https://doi.org/10.1083/jcb.200410090

[184] Rothhaar TL, Grösgen S, Haupenthal VJ, Burg VK, Hundsdörfer B, Mett J, et al. Plasmalogens inhibit APP processing by directly affecting γ -secretase activity in alzheimer's disease. Scientific World Journal. 2012;**2012** https://doi.org/10.1100/2012/141240

[185] Braverman NE, Moser AB. Functions of plasmalogen lipids in health and disease. Biochim Biophys Acta - Mol Basis Dis. 2012;**1822**:1442-1452 https://doi.org/10.1016/j.bbadis.2012.05.008

[186] Alzheimer's Association. 2014 Alzheimer's disease facts and figures. Alzheimers Dement 2014;10:e47-92. https://doi.org/10.1016/j.jalz.2014.02.001.

[187] Brinton RD, Yao J, Yin F, Mack WJ, Cadenas E. Perimenopause as a neurological transition state. Nature Reviews. Endocrinology. 2015;**11**:393-405 https://doi.org/10.1038/nrendo.2015.82

[188] Henderson VW. Cognitive changes after menopause: Influence of estrogen. Clinical Obstetrics and Gynecology. 2008;**51**:618-626 https://doi.org/10.1097/GRF.0b013e318180ba10